Designing to the Strongest Ground Motions

Other Titles of Interest

Site-Specific Ground Motions for Seismic Design of Buildings and Other Structures, by Praveen K. Malhotra (ASCE Press 2022). Author Malhotra provides a step-by-step approach to properly conduct site-specific GMHA and SRA, and informs readers of various resources that are available to perform GMHA and SRA. (ISBN 978-0-7844-1596-2)

Seismic Evaluation and Design of Petrochemical and Other Industrial Facilities, Third Edition, by the Task Committee on Seismic Evaluation and Design of Petrochemical Facilities (ASCE/Energy Division 2020). This report offers practical recommendations regarding the design and safety of new and existing petrochemical facilities during and following an earthquake. (ISBN 978-0-7844-1548-1)

Seismic Loads: Guide to the Seismic Load Provisions of ASCE 7-16, by Finley A. Charney, Thomas F. Heausler, and Justin D. Marshall (ASCE Press 2020). Authors Charney, Heausler, and Marshall provide clear, authoritative explanations of the seismic design provisions contained in Minimum Design Loads and Associated Criteria for Buildings and Other Structures, Standard ASCE/SEI 7-16. (ISBN 978-0-7844-1550-4)

Earthquakes and Engineers: An International History, by Robert K. Reitherman (ASCE Press 2012). Author Reitherman offers the first comprehensive treatment of the engineering techniques devised around the world to address seismic problems and design structures that resist earthquakes. (ISBN 978-0-7844-1071-4)

Primer on Seismic Isolation, Task Committee on Seismic Isolation; edited by Andrew W. Taylor and Takeru Igusa (ASCE/SEI 2004). This primer describes the current state of seismic isolation technology and highlights issues and concerns that are unique to the design of isolated structures. (ISBN 978-0-7844-0751-6)

Designing to the Strongest Ground Motions

Praveen K. Malhotra, Ph.D., P.E.

ASCE PRESS

Published by the American Society of Civil Engineers

Library of Congress Cataloging-in-Publication Data

LCCN 2023037717
Cataloging-in-Publication Data on file with the Library of Congress

Published by American Society of Civil Engineers
1801 Alexander Bell Drive
Reston, Virginia 20191-4382
www.asce.org/bookstore | ascelibrary.org

Errata: Errata, if any, can be found at https://doi.org/10.1061/9780784485194.

ISBN 978-0-7844-1618-1 (print)
ISBN 978-0-7844-8519-4 (PDF)
ISBN 978-0-7844-8520-0 (ePub)

Manufactured in the United States of America.

28 27 26 25 24 23 1 2 3 4 5

Contents

Preface . . . . **ix**

Acknowledgments and Dedication . . . . **xiii**

Abbreviations, Acronyms, and Notation . . . . **xv**

Chapter 1 Strongest Recorded Ground Motions **1**
- 1.1 Introduction 1
- 1.2 Ground Motion Histories 1
- 1.3 Single-Degree-of-Freedom System 6
- 1.4 Deformation Response Spectrum 7
- 1.5 Pseudo-Acceleration Response Spectrum 9
- 1.6 Pseudo-Velocity Response Spectrum 11
- 1.7 Tripartite Plot of Response Spectrum 11
- 1.8 Acceleration–Deformation Response Spectrum 12
- 1.9 Acceleration-Deformation Response Spectra for Various Damping . . . 15
- 1.10 Strongest Recorded Ground Motions 16
- 1.11 Cyclic Demand 18
- 1.12 Indirect Parameters of Ground Motion 20
- 1.13 Summary 21
- References 21

Chapter 2 Comparison with Predicted Ground Motions **23**
- 2.1 Introduction 23
- 2.2 San Francisco, California, Ground Motion 23
- 2.3 Hickman, Arkansas, Ground Motion 25
- 2.4 Discussion 26
- 2.5 Summary 26
- References 27

Chapter 3 Seismic Characteristics of Buildings **29**
3.1 Introduction 29
3.2 Three-Story Steel Special Moment Frame Building 29
3.3 Pushover Curve for 3-Story Building 30
3.4 Capacity Curve for 3-Story Building 35
3.5 Damping Curve for 3-Story Building 37
3.6 Twenty-Story Steel Eccentrically Braced Frame Building 39
3.7 Pushover Curve for 20-Story Building 42
3.8 Capacity Curve for 20-Story Building 46
3.9 Damping Curve for 20-Story Building 48
3.10 Comparison between 3- and 20-Story Buildings 49
3.11 Cyclic Capacity 50
3.12 Summary 52
References 53

Chapter 4 Seismic Analysis of Buildings **55**
4.1 Introduction 55
4.2 Demand Curve for 3-Story Building 55
4.3 Nonlinear-Static Analysis of 3-Story Building 58
4.4 Nonlinear-Dynamic Analysis of 3-Story Building 59
4.5 Damage Suffered by 3-Story Building 62
4.6 Demand Curve for 20-Story Building 62
4.7 Nonlinear-Static Analysis of 20-Story Building 63
4.8 Nonlinear-Dynamic Analysis of 20-Story Building 66
4.9 Damage Suffered by 20-Story Building 67
4.10 Summary 68
References 69

Chapter 5 Collapse-Inducing Ground Motions **71**
5.1 Introduction 71
5.2 Weakest Collapse-Inducing Ground Motion 71
5.3 Effect of Deformability on Collapse-Inducing Ground Motion 78
5.4 Effect of Hysteresis on Collapse-Inducing Ground Motion 79
5.5 Summary 81
References 81

Chapter 6 Response of Buildings to Strongest Ground Motions **83**
6.1 Introduction 83
6.2 Northridge, Tarzana Ground Motion 83
6.3 Loma Prieta, Los Gatos Presentation Center Ground Motion 87
6.4 Chi-Chi, Taiwan, TCU065 Ground Motion 90
6.5 Northridge, Sylmar Ground Motion 94
6.6 Chi-Chi, Taiwan, CHY080 Ground Motion 97

6.7 Kobe, Takatori Ground Motion 101
6.8 Chi-Chi, Taiwan, TCU084 Ground Motion 104
6.9 Nurdagi, Türkiye, TK3123 Ground Motion 108
6.10 Chi-Chi, Taiwan, TCU052 Ground Motion 111
6.11 Niigata, Japan, NIG019 Ground Motion 114
6.12 Chi-Chi, Taiwan, TCU068 Ground Motion 118
6.13 Damage from Aftershocks 121
6.14 Summary 123
References 124

Chapter 7 Collapse-Proof Seismic Design 125
7.1 Introduction 125
7.2 Envelope Response Spectra 125
7.3 Preliminary Design 128
7.4 Capacity and Damping Curves 129
7.5 Soil–Structure Interaction 131
7.6 Final Design 132
7.7 Summary 132
References 133

Chapter 8 R, C_d, and Ω_0 Factors 135
8.1 Introduction 135
8.2 Three-Story Steel Special Moment Frame Building 136
8.3 Twenty-Story Steel Eccentrically Braced Frame Building 140
8.4 Summary 143
References 144

Chapter 9 Concluding Remarks 145

Appendix Damping of Hysteretic System 147
References 149

Index 151

About the Author 157

Preface

Future ground motions are highly uncertain, but the response of structures to future ground motions need not be uncertain. Through science and engineering, uncertainty can be reduced to a negligible level.

It is generally assumed that human-made structures cannot withstand the strongest ground motions; therefore, the risk of collapse is unavoidable. Seismic design of ordinary buildings in the United States accepts 1% probability of collapse in 50 years (Luco et al. 2007, ASCE 7-22). However, it is very difficult to accept collapses of the following structures:

- Hospital that is needed to treat the sick and the injured after an earthquake.
- School that doubles as an emergency shelter.
- Dam whose collapse may flood an entire city.
- Nuclear containment structure whose breach may expose a large population to radiation.
- Heavy structures in urban areas whose simultaneous collapse can kill thousands.

Earthquakes provide zero to negligible warning, and they can collapse numerous structures at the same time. How many buildings can be allowed to simultaneously collapse in a major earthquake has not been discussed by Luco et al. (2007) or anyone else (Malhotra 2009). Earthquake scenarios for insurance and emergency-planning purposes predict numerous collapses during major events in populated areas (CEA 2019, Jones et al. 2008, USGS 2023). Due to the high cost of repair and rebuilding, the earthquake risk needs to be insured (transferred) for a certain cost (premium). The insurance companies protect themselves from aggregate loss (resulting from simultaneous damage or collapse of multiple structures) by limiting their exposure. In high-seismic urban areas, the cost of insurance is so high that most structures are either uninsured or inadequately insured against earthquakes.

Two questions arise:

1. Are there enough data to determine the strongest ground motions?
2. Can structures be designed to withstand the strongest ground motions?

In response to the first question, seismic instruments have been deployed to measure and record ground motions since 1930. Thousands of recording stations currently exist in populated, high-seismic areas. Approximately 5,000 stations, with an average recording time of 40 years, have provided 5,000 × 40 = 200,000 instrument-year of recorded data. Strongest ground motions determined from these data have 50/200,000 × 100 = 0.025% probability of exceedance in 50 years at any specific site. This is 1/40th the collapse probability accepted by the current seismic design in the United States. Therefore, designing structures to withstand the strongest recorded ground motions will reduce the collapse probability to a negligible level. The second question is examined in this book.

Chapter 1 discusses ground motion characteristics. It focuses on engineering rather than seismological characteristics. Eleven strongest ground motions are selected from worldwide recordings.

Chapter 2 compares the strongest recorded ground motions with those predicted by the probabilistic seismic hazard analysis (PSHA).

Chapter 3 discusses seismic characteristics of buildings. Two buildings are described in detail: (1) 3-story steel special moment frame (SMF) building and (2) 20-story steel eccentrically braced frame (EBF) building.

Chapter 4 describes methods of analysis that are physics based and code independent. They are applicable anywhere in the world. Cumulative damage from cycles of various amplitudes is also considered.

Chapter 5 discusses the characteristics of collapse-inducing ground motions. Weakest ground motions that can collapse brittle, ductile, and ductile-hysteretic structures are discussed.

Chapter 6 discusses the performance of 3-story steel SMF and 20-story steel EBF buildings to each of the 11 strongest ground motions selected in Chapter 1. The effect of aftershocks is also discussed.

Chapter 7 describes the necessary steps for a collapse-proof seismic design. The example of a buckling-restrained braced frame (BRBF) building is discussed. The effect of soil–structure interaction (SSI) is also discussed.

Chapter 8 explores the use of response modification factor R, deflection amplification factor C_d, and overstrength factor Ω_0 in collapse-proof seismic design.

Chapter 9 concludes the book with a summary.

Praveen K. Malhotra
Sharon, Massachusetts, USA

References

ASCE. 2022. *Minimum design loads and associated criteria for buildings and other structures*, ASCE 7-22. Reston, VA: ASCE.

CEA (California Earthquake Authority). 2019. "What to expect from an earthquake along the Hayward Fault." Sacramento, CA: CEA. https://www.earthquakeauthority.com/Blog/2019/hayward-fault-earthquake-prediction.

Jones, L. M., R. Bernknopf, D. Cox, J. Goltz, et al. 2008. "The ShakeOut scenario." Earthquake Hazards Program, Reston, VA: US Geological Survey. https://pubs.usgs.gov/of/2008/1150/.

Luco, N., B. R. Ellingwood, R. O. Hamburger, J. D. Hooper, et al. 2007. "Risk-targeted vs. current seismic design maps for the conterminous United States." In *Proc., SEAOC 76th Annual Convention*, Sacramento, CA: Structural Engineers Association of California.

Malhotra, P. K. 2009. "Should building codes target societal risk?" *Earthquake Spectra* 25 (1): 213–217.

USGS (US Geological Survey). 2023. "Hayward Fault scenarios." Reston, VA: USGS. https://earthquake.usgs.gov/education/shakingsimulations/hayward/.

Acknowledgments

Professor Cheng Chen of San Francisco State University, Dr. Kun David Li of WSP, and an anonymous reviewer provided numerous helpful suggestions to improve this book.

Natalie Webster, Michie Gluck, and Donna Dickert of ASCE Press are thanked for their professionalism and willingness to help during various stages of this book's publication.

My loving wife, Romee, was always there to encourage and support me.

Dedication

I dedicate this book to the memory of Matthew Cowden and Thomas Cooper, outstanding engineers, who asked apt questions.

Abbreviations, Acronyms, and Notation

ADRS	Acceleration-deformation response spectrum
A_s	Shear area of beam for calculating V_p
ASCE	American Society of Civil Engineers
BRB	Buckling-restrained brace
BRBF	Buckling-restrained braced frame
CDRS	Cyclic-demand response spectrum
CG	Center of gravity
D	Deflection of CG
D_e	Maximum elastic deflection of CG
D_{ult}	Deflection of CG at collapse
e	Length of link beam in EBF
EBF	Eccentrically braced frame
ED	Energy demand
EOM	Equation of motion
F	Lateral force on CG of the building
F_y	Minimum yield stress of steel
F_{ye}	Expected yield stress of steel
g	386 in./s^2 = acceleration due to gravity
H	Total height of the building
H_{CG}	Height of CG of the building
k	Stiffness of SDOF system or elastic stiffness of building
L	Horizontal distance between centerlines of columns
LGPC	Los Gatos Presentation Center
m	Mass of the SDOF system
M	Magnitude of the earthquake
M	Mass of the building
MCE$_R$	Maximum considered earthquake risk-targeted ground motions
M_p	Plastic moment capacity of beam cross section
N_{eq}	Equivalent number of cycles
PD	Peak deformation
PF	Peak force
PGA	Peak horizontal ground acceleration

PGD	Peak horizontal ground displacement
PGV	Peak horizontal ground velocity
PGV_n	Normalized velocity
PPA	Peak pseudo-acceleration
PPV	Peak pseudo-velocity
PSE	Peak strain energy
R_{jb}	Minimum distance from surface projection of the fault; aka Joyner-Boore distance
RotD50	Median (50th percentile) of maxima in all horizontal directions
RotD100	Maximum of maxima in all horizontal directions
SDOF	Single-degree-of-freedom system
SF	Safety factor
SMF	Special moment frame
SSI	Soil-structure interaction
ST	Seismic toughness
T	Natural period of structure
TAPS	Trans-Alaskan Pipeline Pump Station
T_c	Central period of ground motion
T_{cg}	Energy centroid of ground motion
T_{eq}	Effective period of the building at equilibrium
T_u	Longest usable period of the ground motion
T_{ult}	Effective period of the building at collapse
USGS	United States Geological Survey
V_p	Plastic shear capacity of beam cross section
W	Weight of building
Z_p	Plastic modulus of beam cross section for calculating M_p
α	Post-/pre-yield stiffness
β	Ratio between the areas of realistic and ideal hysteresis loops
ζ	Viscous damping ratio
ζ_e	Viscous damping ratio during elastic deformation
μ	Ductility ratio = maximum/elastic deflection
θ	Total (elastic + plastic) drift
θ_p	Plastic drift ratio
θ_p^b	Plastic hinge rotation in beam due to flexural yielding
θ_{ult}	Total (elastic + plastic) drift ratio at collapse
γ	Total (elastic + plastic) shear rotation in link beam
γ_p	Plastic shear rotation in link beam

CHAPTER 1

Strongest Recorded Ground Motions

1.1 Introduction

During an earthquake, the ground shakes in all directions. The inertia of structures prevents them from moving in unison with the ground. As a result, structures deform and develop stresses. Every structure has a certain capacity to deform. If the deformation demand exceeds the deformation capacity, the structure collapses. This chapter deals with the deformation demand imposed by the ground motion. Chapter 3 deals with the deformation capacity (deformability) of the structure.

1.2 Ground Motion Histories

Ground motions for engineering applications are measured by accelerometers (acceleration sensors) oriented in three orthogonal directions – usually east, north, and upward. Recorded accelerations are corrected for instrument response, baseline drift, and perceived "noise" (Hudson 1979).

Figure 1-1 shows the corrected acceleration histories of ground motion recorded at the Trans-Alaskan Pipeline Pump Station 10 (TAPS) during the 2002 magnitude **M** 7.9 Denali Earthquake in Alaska (COSMOS 2022). The TAPS 10 station is owned and maintained by the Alyeska Pipeline Service Company. The recorded histories were corrected by applying a high-pass filter with a cutoff frequency of 0.1 Hz. The lowest usable frequency of the corrected ground motion is 0.125 Hz, or the longest usable period is $T_u = 1/0.125 = 8$ s. As is common, in Figure 1-1, the accelerations are expressed in units of $g = 386$ in./s^2 = acceleration due to gravity. The horizontal motion of the ground is of greater interest to engineers than the vertical motion because most structures are designed to resist vertical (gravity) loads with high margin of safety. Apparent change in the gravity load due to vertical ground motion does not pose a problem for most engineered structures, but there can be exceptions. This book focuses only on the horizontal ground motions.

In Figure 1-1, the maximum accelerations in the east and north directions are $0.307g$ and $0.358g$, respectively. The peaks in east and north directions do not occur at the same time. It is likely that the maximum-horizontal acceleration is not in the

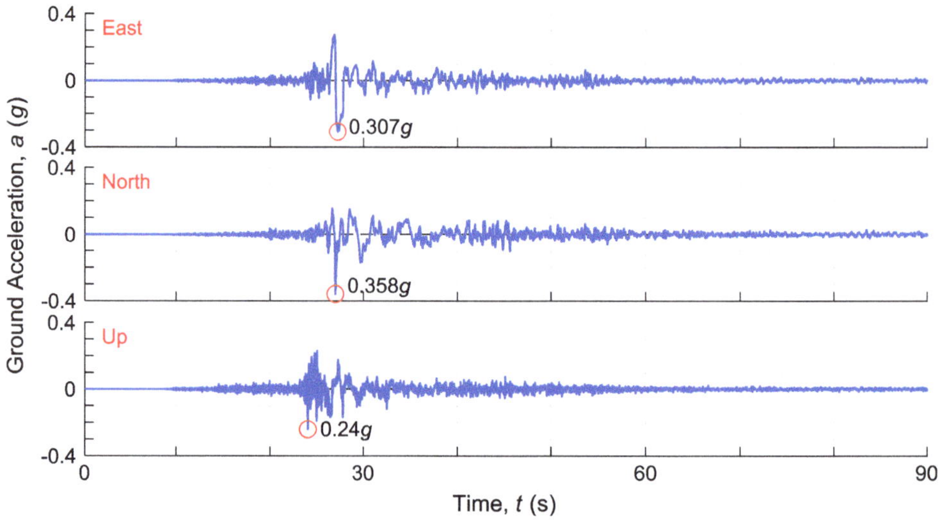

Figure 1-1. Corrected acceleration histories at TAPS 10 station during the 2002 magnitude **M** 7.9 Denali Earthquake in Alaska.

east or north direction but in some other direction in the horizontal plane. To determine the maximum-horizontal acceleration, the resultant accelerations are calculated at each instant from the accelerations in the east and north directions:

$$a_r = \sqrt{a_e^2 + a_n^2} \tag{1-1}$$

where a_e, a_n, and a_r are the instantaneous values of east, north, and resultant horizontal accelerations, respectively. The instantaneous direction of the resultant horizontal acceleration (clockwise from north) is

$$\phi = \tan^{-1} a_e / a_n \tag{1-2}$$

Figure 1-2 shows a plot of the resultant horizontal accelerations of the TAPS ground motion. The maximum-horizontal ground acceleration is $PGA = 0.359\,g$, and it occurs in the $\phi = 187°$ direction. Notice in Figure 1-2 that the maximum acceleration is different in different horizontal directions. The median (50th percentile) of maxima in all horizontal directions is denoted as RotD50 (Boore et al. 2006). The maximum of maxima in all horizontal directions is denoted as RotD100. Both RotD50 and RotD100 are independent of the orientation of seismic instruments (Boore et al. 2006). Seismologists prefer RotD50, but engineers require RotD100 (Baker and Cornell 2006). In this book, RotD100 (maximum-horizontal) is used for all ground motion parameters.

The ground accelerations in Figure 1-1 are numerically integrated to obtain ground velocities (Hudson 1979). Any trends in the velocity histories are removed to

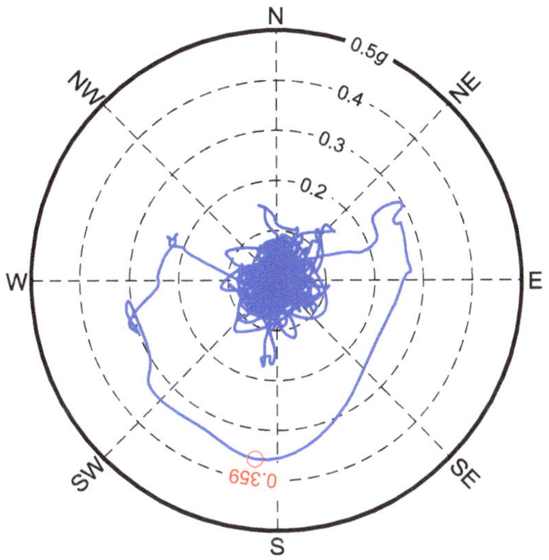

Figure 1-2. Resultant horizontal accelerations during the TAPS ground motion.

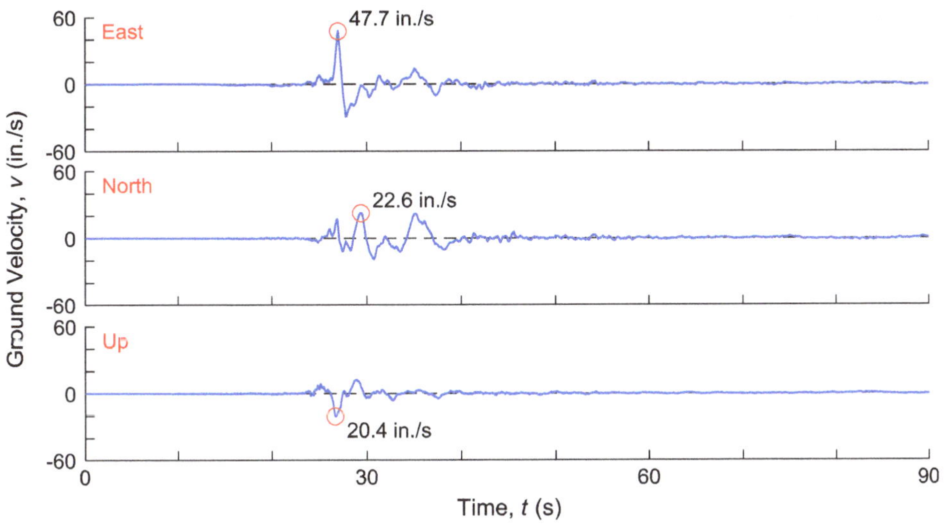

Figure 1-3. Velocity histories in three orthogonal directions during the TAPS ground motion.

ensure that the velocities are zero at the end. Figure 1-3 shows corrected velocity histories in the east, north, and upward directions. The resultant horizontal velocities are calculated in the same manner as the resultant horizontal accelerations. Figure 1-4 shows resultant horizontal velocities of the TAPS ground motion. The maximum-horizontal ground velocity is $PGV = 49.1$ in./s.

The ground velocities in Figure 1-3 are numerically integrated to obtain ground displacements (Hudson 1979). Any trends in the displacement histories that are not

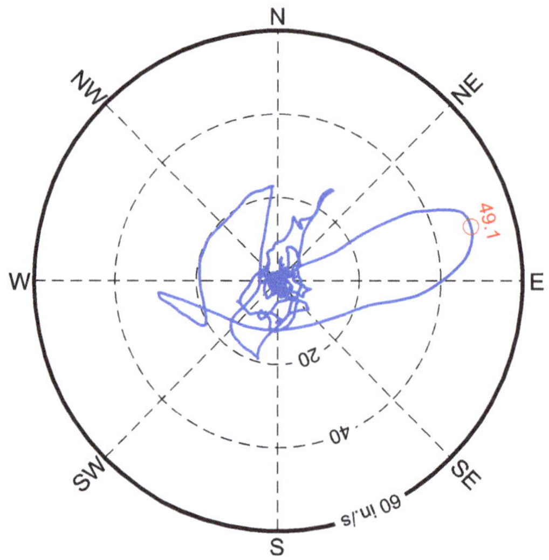

Figure 1-4. Resultant horizontal velocities during the TAPS ground motion.

consistent with seismic wave theory are removed through filtering or trend removal. Figure 1-5 shows corrected displacement histories in the east, north, and upward directions. The displacements derived from acceleration measurements may not be very accurate because slow movements of the ground can produce large displacements without producing measurable accelerations. However, slow movements of

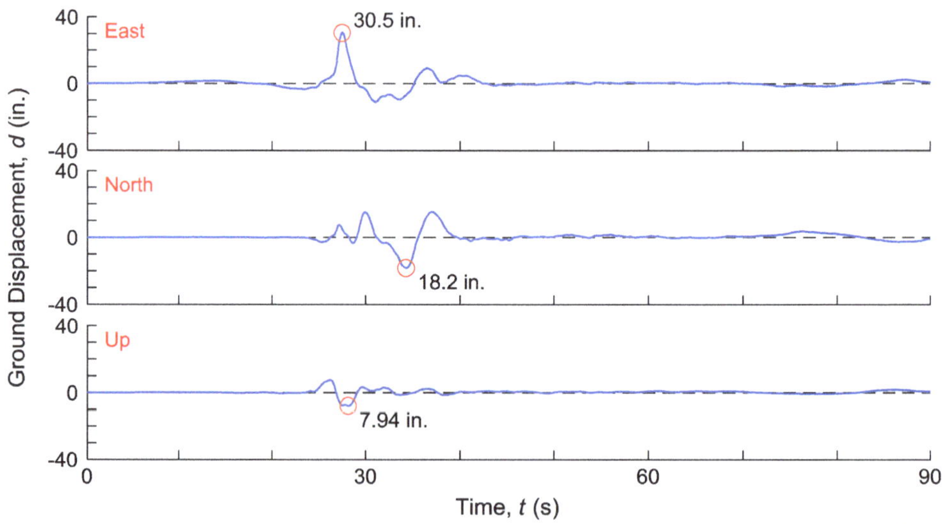

Figure 1-5. Displacement histories in three orthogonal directions during the TAPS ground motion.

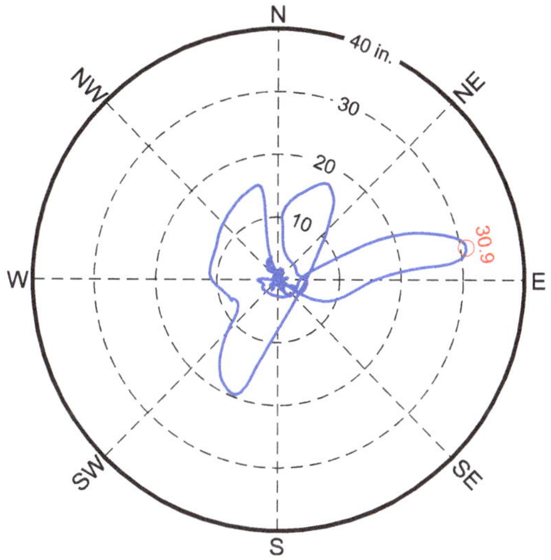

Figure 1-6. Resultant horizontal displacements during the TAPS ground motion.

the ground do not induce dynamic response in most structures because most structures do not respond to very low frequencies associated with slow movements.

The resultant horizontal displacements are calculated in the same manner as the resultant horizontal accelerations and velocities. Figure 1-6 shows resultant horizontal displacements of the TAPS ground motion. The maximum-horizontal ground displacement is $PGD = 30.9$ in. Note that in Figures 1-2, 1-4, and 1-6, directions of maximum horizontal (RotD100) acceleration, velocity, and displacement are not the same.

The central period of the ground motion is defined (Malhotra 2021) as

$$T_c = 2\pi\sqrt{\frac{PGD}{PGA}} \tag{1-3}$$

Ground motions with $T_c > 2$ s are low-frequency and those with $T_c < 0.5$ s are high-frequency (Malhotra 2021). Substituting $PGA = 0.359g = 139\ \text{in./s}^2$ and $PGD = 30.9$ in. into Equation (1-3), $T_c = 3$ s is obtained. Since $T_c > 2$ s, TAPS ground motion is low frequency. The normalized velocity of a ground motion is defined (Malhotra 2021) as

$$PGV_n = \frac{PGV}{\sqrt{PGA \cdot PGD}} \tag{1-4}$$

Ground motions with $PGV_n > 0.75$ are narrow banded, and those with $PGV_n < 0.45$ are broad banded. Substituting $PGA = 0.359g = 139\ \text{in./s}^2$, $PGV = 49.1$ in./s, and $PGD = 30.9$ in. into Equation (1-4), $PGV_n = 0.75$ is obtained. Hence, the TAPS ground motion is medium to narrow banded.

PGA, *PGV*, and *PGD* define the amplitude of high, medium, and low frequencies in the ground motion (Malhotra 2021). A more complete (frequency-by-frequency) description of the ground motion is provided by its response spectrum (Newmark and Hall 1982, Clough and Penzien 2003, Chopra 2017), but it takes some additional effort to generate the response spectrum of a ground motion. To describe the response spectrum, a simple mathematical model of the structure is discussed first.

1.3 Single-Degree-of-Freedom System

A linear-elastic structure can be represented by one or more single-degree-of-freedom (SDOF) systems of the type shown in Figure 1-7; m is the mass, k is the stiffness, and c is the viscous damping constant. In Figure 1-7, only the spring k and the damper c deform; all other elements are rigid. The undamped period T of the system is given (e.g., Chopra 2017) by the following expression:

$$T = 2\pi\sqrt{\frac{m}{k}} \tag{1-5}$$

The damping ratio ζ is defined (e.g., Chopra 2017) as

$$\zeta = \frac{c}{2\sqrt{km}} \tag{1-6}$$

The mass m in Figure 1-7 is displaced 1 in. and set free. The support on the right side of Figure 1-7 is kept fixed. Figure 1-8 shows the free-vibration responses of undamped ($\zeta = 0$), 5% damped ($\zeta = 0.05$), and critically damped ($\zeta = 1$) systems of period $T = 1$ s. The higher the damping, the greater the rate of energy dissipation and the faster is the decay of amplitude. A linear-elastic SDOF system is completely defined by three parameters: (a) m, k, and c, or (b) m, T, and ζ.

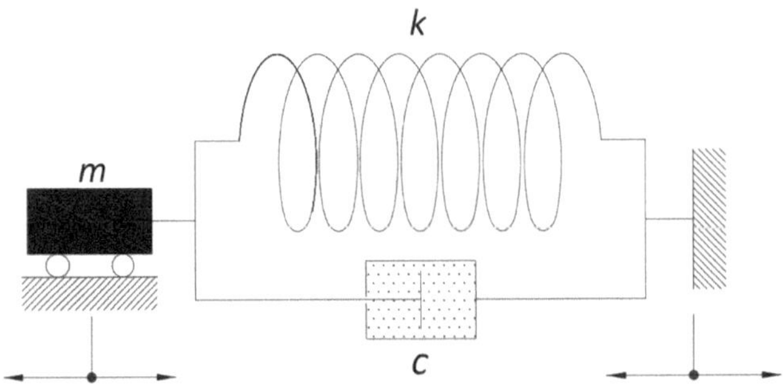

Figure 1-7. Single-degree-of-freedom model of a structure.

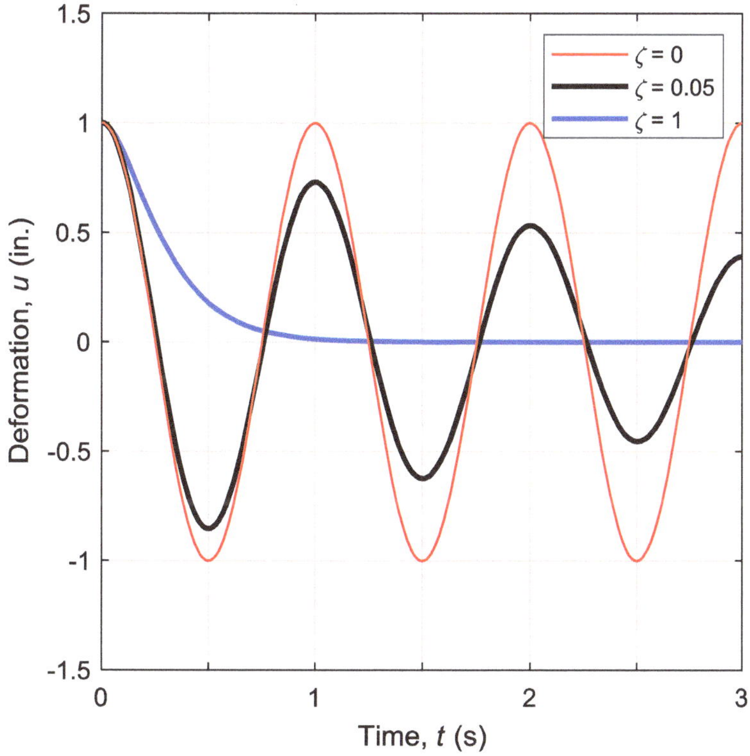

Figure 1-8. Free-vibration responses of single-degree-of-freedom systems of period $T = 1$ s and various damping.

1.4 Deformation Response Spectrum

Consider an SDOF system of period $T = 1$ s and damping $\zeta = 0.05$ (5% of critical). The support on the right side of Figure 1-7 is moved with the east-direction ground motion shown in Figures 1-1, 1-3, and 1-5. The equation of motion (EOM) is solved by the method described by Malhotra (2001) to compute the east-direction deformation history shown in the upper part of Figure 1-9. The deformation u is the movement of the mass m with respect to the support. Next, the support is moved with the north-direction ground motion shown in Figures 1-1, 1-3, and 1-5. The EOM is again solved by the method described by Malhotra (2001) to compute the north-direction deformation history shown in the lower part of Figure 1-9. The peak deformations PD in the east and north directions are 9.41 and 6.18 in., respectively. The amplitude and direction of resultant horizontal deformation are computed at various instants from the instantaneous deformations in the east and north directions. Figure 1-10 shows a plot of the resultant horizontal deformations of the SDOF system during the TAPS ground motion. The maximum-horizontal peak deformation of 1 s period, 5% damped system is $PD(1\ \text{s}) = 9.96$ in., and it occurs in $\phi = 70°$ clockwise from north or N70E direction.

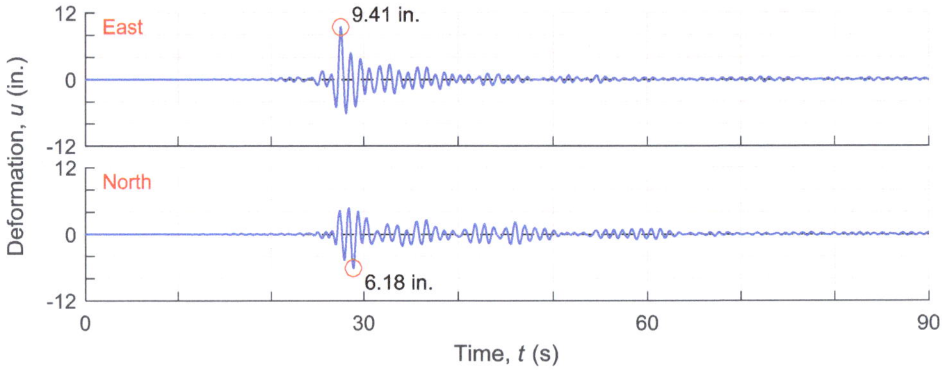

Figure 1-9. Deformations of the single-degree-of-freedom system of 1 s period, and 5% damping subjected to horizontal TAPS ground motion shown in Figures 1-1 to 1-6.

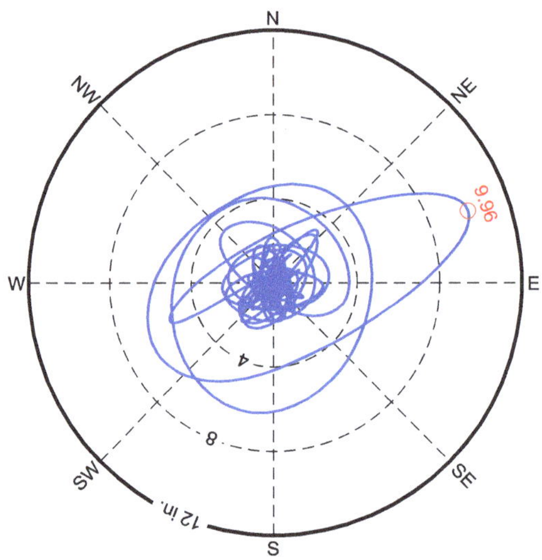

Figure 1-10. Resultant horizontal deformations of the single-degree-of-freedom system of 1 s period, and 5% damping, subjected to TAPS ground motion shown in Figures 1-1 to 1-6.

Maximum-horizontal peak deformation *PD* is similarly computed for other values of natural period T. The deformation response spectrum of a ground motion is a plot between the natural period T and peak deformation *PD*, for a fixed value of damping ζ. Figure 1-11 shows the 5% damping, maximum-horizontal deformation response spectrum of the TAPS ground motion shown in Figures 1-1 to 1-6. Due to the type of filter applied in processing the TAPS ground motion, the longest usable period is $T_u = 8$ s. Therefore, the less-reliable portion of the response spectrum, beyond the 8 s period, is shown in light-blue color. All analyses in this book are carried out using the entire response spectrum up to the 100 s period. The reader is

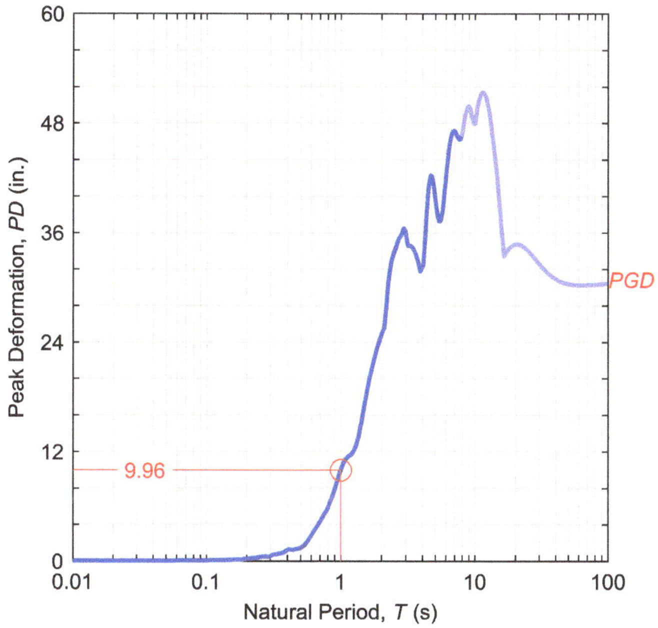

Figure 1-11. Five-percent damping, maximum-horizontal deformation response spectrum of TAPS ground motion shown in Figures 1-1 to 1-6.

cautioned whenever the results of the analysis are affected by the less-reliable portion of the response spectrum.

1.5 Pseudo-Acceleration Response Spectrum

The peak force is the product of peak deformation and stiffness:

$$PF = k \cdot PD \tag{1-7}$$

The peak force PF is also known as the spring force; it does not include the force due to damping. With the help of Equation (1-5), the stiffness k is expressed in terms of m and T to yield the following expression for the peak force:

$$PF = m \cdot \left(\frac{2\pi}{T}\right)^2 \cdot PD \tag{1-8}$$

The multiplier of mass m in Equation (1-8) has the units of acceleration, but it is approximately equal to the "true" peak acceleration of the mass. Therefore, it is simply called the peak pseudo-acceleration PPA, defined as follows:

$$PPA = \left(\frac{2\pi}{T}\right)^2 \cdot PD \tag{1-9}$$

The difference between PPA and "true" acceleration is due to the damping force. Engineers are mostly interested in the spring force. The expression for the peak force in the spring [Equation (1-8)] can be rewritten as

$$PF = m \cdot PPA \tag{1-10}$$

Note that the "true" force in the spring is obtained by multiplying the mass by the pseudo-acceleration. A plot of PPA versus T is known as the pseudo-acceleration response spectrum. A pseudo-acceleration response spectrum can be generated from a deformation response spectrum simply by using Equation (1-9). For example, $PD(1\text{ s}) = 9.96\text{ in.}$ Therefore, $PPA(1\text{ s}) = (2\pi/1)^2 \times 9.96 = 393\text{ in./s}^2 = 1.02g$. PPA can be similarly calculated for other values of period. Figure 1-12 shows the 5% damping, maximum-horizontal pseudo-acceleration response spectrum of the TAPS ground motion shown in Figures 1-1 to 1-6. The response spectrum beyond the longest usable period of $T_u = 8\text{ s}$ is shown in light-blue color. Note that maximum values of PD and PPA in Figures 1-11 and 1-12, respectively, occur at different periods. Only for very narrow banded ground motions do the maximum values of PD and PPA occur at the same period.

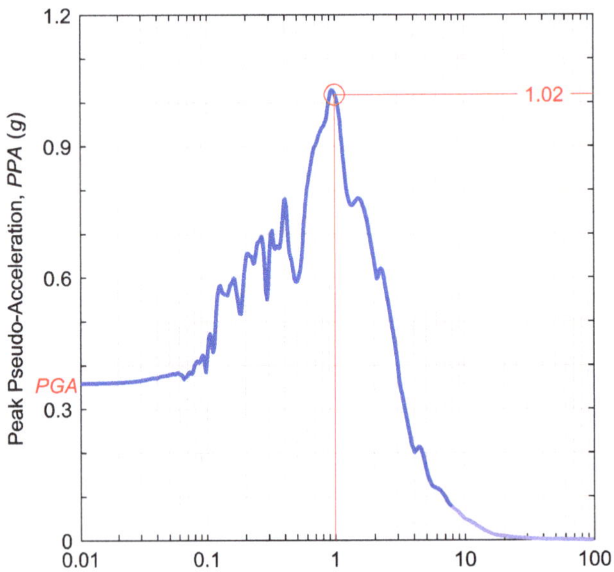

Figure 1-12. Five-percent damping, maximum-horizontal pseudo-acceleration response spectrum of TAPS ground motion shown in Figures 1-1 to 1-6.

1.6 Pseudo-Velocity Response Spectrum

The peak strain energy is

$$PSE = \frac{1}{2}k \cdot PD^2 \tag{1-11}$$

With the help of Equation (1-5), the stiffness k can be expressed in terms of m and T to yield the following expression for the peak strain energy:

$$PSE = \frac{1}{2}m \cdot \left(\frac{2\pi}{T} \cdot PD\right)^2 \tag{1-12}$$

The term in parentheses in the previous expression has the units of velocity, but it is not the true velocity of the mass. Therefore, it is simply called the peak pseudo-velocity, PPV, defined as

$$PPV = \frac{2\pi}{T} \cdot PD \tag{1-13}$$

The expression for the peak strain energy [Equation (1-12)] can be rewritten as

$$PSE = \frac{1}{2}m \cdot PPV^2 \tag{1-14}$$

Note that the "true" strain energy is obtained by multiplying half the mass by the square of the "pseudo-velocity."

A plot of PPV versus T is known as the pseudo-velocity response spectrum. The pseudo-velocity response spectrum can be generated from the deformation response spectrum simply by using Equation (1-13). For example, $PD(1\,\text{s}) = 9.96\,\text{in.}$ Therefore, $PPV(1\,\text{s}) = (2\pi/1) \times 9.96 = 62.6\,\text{in./s}$. PPV can be similarly calculated for other periods. Figure 1-13 shows the 5% damping, maximum-horizontal pseudo-velocity response spectrum of the TAPS ground motion. The response spectrum beyond the longest usable period of $T_u = 8\,\text{s}$ is shown in light-blue color.

1.7 Tripartite Plot of Response Spectrum

The PPA, PPV, and PD response spectra can be thought of as the plots of high, intermediate, and low frequencies in the ground motion. They are related to each other as follows:

$$PPA \cdot \frac{T}{2\pi} = PPV = \frac{2\pi}{T} \cdot PD \tag{1-15}$$

The above relationship makes it possible to display all three response spectra by a single curve in a tripartite format. Figure 1-14 displays the tripartite plot of the 5% damping, maximum-horizontal response spectrum of TAPS ground motion shown

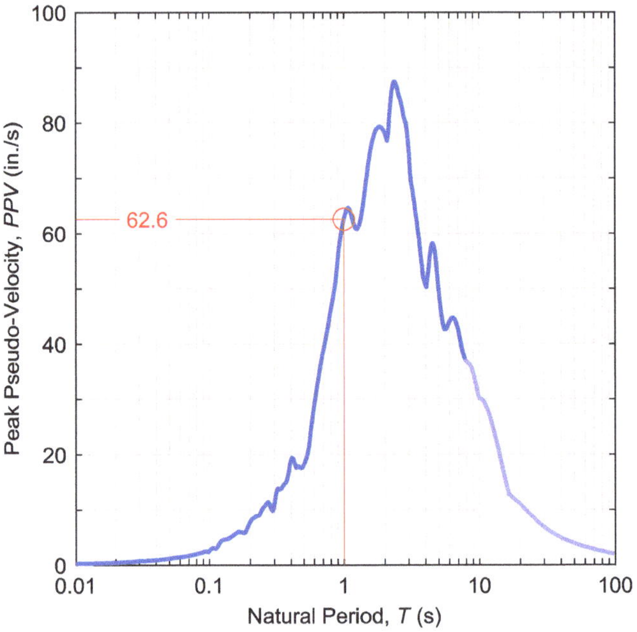

Figure 1-13. Five-percent damping, maximum-horizontal pseudo-velocity response spectrum of TAPS ground motion shown in Figures 1-1 to 1-6.

in Figures 1-1 to 1-6. In this plot, PPV is read along the vertical axis, period T is read along the horizontal axis, PPA is read along the −45 degrees (counterclockwise) axis, and PD is read along the +45 degrees (clockwise) axis. For reference, PGA, PGV, and PGD are shown by three straight lines. Once again, the response spectrum beyond the longest usable period of $T_u = 8\,\text{s}$ is shown in light-blue color. The tripartite plot clearly shows that PPA approaches PGA at very short periods and PD approaches PGD at very long periods. The tripartite plot is a very concise and elegant way of displaying the response spectrum of a ground motion. The method of generating a tripartite plot is discussed in Section 2.5 of Malhotra (2022).

1.8 Acceleration–Deformation Response Spectrum

The tripartite response spectrum of Figure 1-14 can be rotated clockwise by 45 degrees such that PPA is along the vertical axis and PD is along the horizontal axis, as shown in Figure 1-15. This is known as the acceleration–deformation response spectrum (ADRS) on a logarithmic scale. Period T is indicated by parallel diagonal lines on a logarithmic ADRS. Figure 1-16 shows a linear plot of the ADRS. Period T is indicated by radial tick marks (passing through the origin) on a linear ADRS.

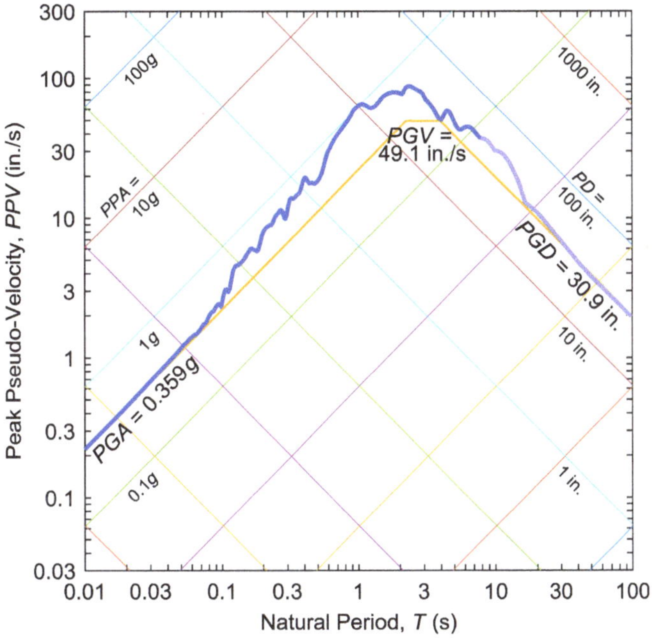

Figure 1-14. Tripartite plot of 5% damping, maximum-horizontal response spectrum of TAPS ground motion shown in Figures 1-1 to 1-6.

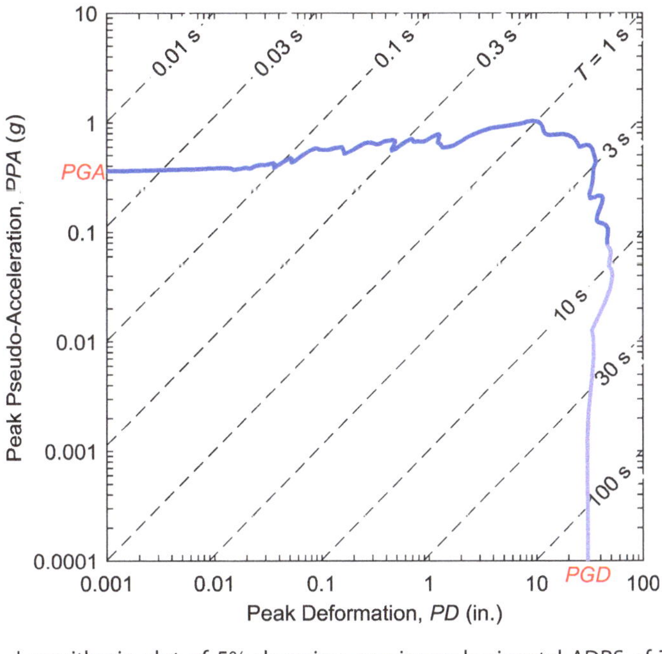

Figure 1-15. Logarithmic plot of 5% damping, maximum-horizontal ADRS of TAPS ground motion shown in Figures 1-1 to 1-6.

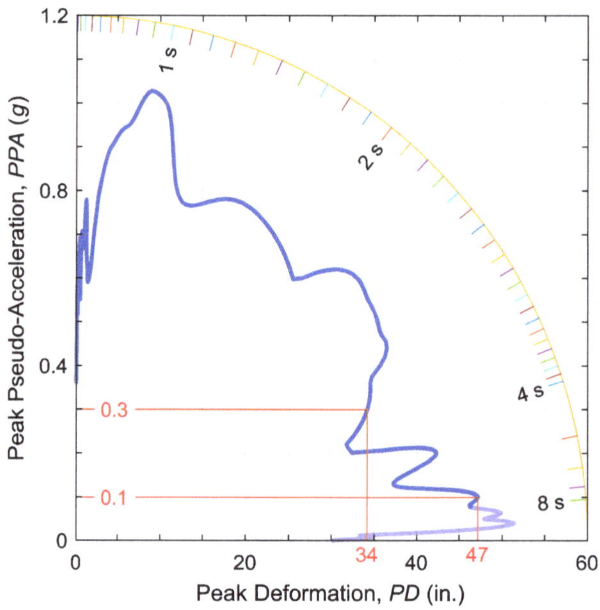

Figure 1-16. Linear plot of 5% damping, maximum-horizontal ADRS of TAPS ground motion shown in Figures 1-1 to 1-6.

ADRS is helpful in estimating the response of nonlinear systems. When shaken strongly, a "ductile" structure yields upon reaching its maximum strength. After yielding, the structure no longer behaves in a linear fashion. Its period and damping change due to yielding. For a nonlinear system, normalized strength rather than period is used to estimate the deformation, where normalized strength = lateral strength divided by mass = peak pseudo-acceleration. As seen in Figure 1-16, a 5% damped structure of normalized strength $0.1g$ is expected to deform 47 in., whereas a structure of normalized strength $0.3g$ is expected to deform 34 in. In general, an increase in normalized strength reduces the deformation of a structure.

In Figures 1-15 and 1-16, the less reliable portion of the response spectrum (shown in light-blue color) corresponds to normalized lateral strength of <0.08 (8% weight of structure). Engineered structures rarely have such low strengths. Therefore, high-pass filter applied on TAPS ground motion does not limit its ability to predict the response of engineered structures.

The area enclosed within the ADRS is known as energy demand, *ED* (Malhotra 2021). *ED* has units of energy per unit mass. *ED* is a measure of deformation demand on structures of various strengths. For TAPS ground motion, energy demand is $ED = 10{,}800\,(\text{in./s})^2 = 6.98\,(\text{m/s})^2$, and energy centroid is $T_{cg} = 2.2$ s, as shown in Figure 1-17. Note that the central period T_c [Equation (1-3)] approximates the energy centroid T_{cg}. In calculating *ED*, the response spectrum beyond the longest usable period of $T_u = 8$ s is used without any modification. Alternatively, the response spectrum beyond T_u can be replaced by a smooth response spectrum

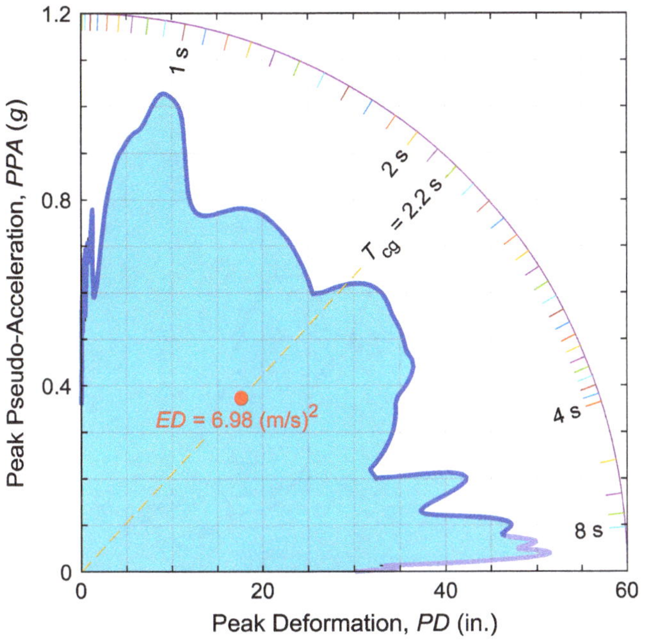

Figure 1-17. Energy demand (*ED*) and energy centroid (T_{cg}) of horizontal TAPS ground motion shown in Figures 1-1 to 1-6.

generated from *PGA*, *PGV*, and *PGD* (Malhotra 2021). While *PGA* and *PGV* are accurately known, *PGD* is not accurately known. *PGD* can be chosen such that the smooth response spectrum matches the actual response spectrum at $T = T_u$. If this procedure is used to modify the response spectrum beyond T_u, *ED* would change from 6.98 to 6.85 $(m/s)^2$, which is a 2% reduction in *ED*. Throughout this book, the response spectrum beyond T_u is used without any modification.

ED can be thought of as the available energy that could be transmitted to a structure. But only a fraction of this energy is actually transmitted to a structure, as will be seen later.

1.9 Acceleration-Deformation Response Spectra for Various Damping

The effect of yielding on period and damping of a structure is discussed in Chapter 3. The 5% damping in prescriptive seismic design is somewhat arbitrarily selected. The actual damping of a structure can be significantly different from 5% of critical. Figure 1-18 shows maximum-horizontal ADRS of TAPS ground motion for various values of damping. For a structure of normalized strength 0.3*g*, increase in damping from 5% to 20% reduces the deformation from 34 to 21 in.

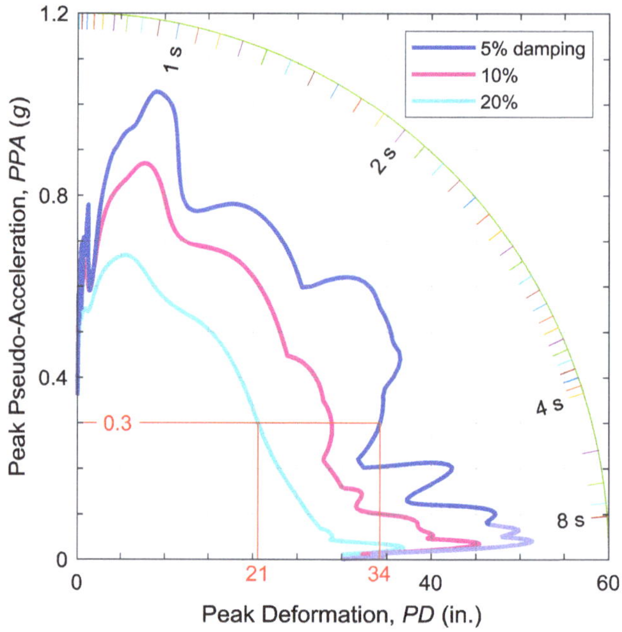

Figure 1-18. Effect of damping on maximum horizontal ADRS of TAPS ground motion shown in Figures 1-1 to 1-6.

1.10 Strongest Recorded Ground Motions

Ground motions with higher energy demand produce greater deformation in structures of various strengths. Therefore, ED determines the damage caused by a ground motion to a broad range of structures. But ED is not the only parameter that determines the damage potential of a ground motion. Energy centroid T_{cg} and normalized velocity PGV_n are other important parameters. Ground motions with shorter T_{cg} are considered "high frequency." Such ground motions are damaging to stiff and brittle structures such as low-rise unreinforced masonry buildings. Ground motions with longer T_{cg} are considered "low frequency". Such ground motions are damaging to weaker but deformable structures such as high-rise buildings. For ground motions with higher PGV_n, energy is concentrated within a narrow range of frequencies. Such ground motions are highly damaging to specific structures. While ED defines the size of the response spectrum, T_{cg} and PGV_n define its shape, but the response spectrum is too irregular to be completely defined by these three parameters.

The energy demand ED correlates well with the square of peak ground velocity, PGV. Figure 1-19 shows a plot between ED and PGV for 3,515 ground motions recorded worldwide. On average,

$$ED \approx 8\,PGV^2 \qquad (1\text{-}16)$$

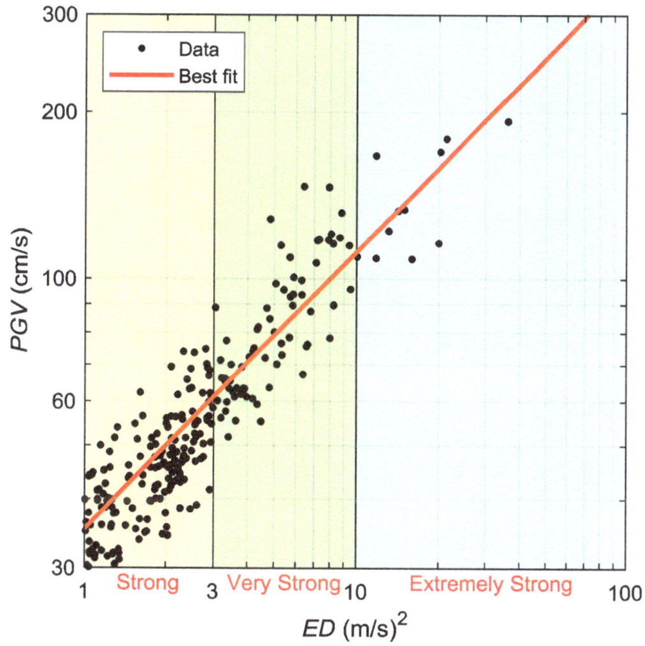

Figure 1-19. *ED* and *PGV* of 3,515 recorded ground motions.

The correlation between ED and PGV^2 is 94%. Therefore, either ED or PGV can be used to identify the strongest ground motions. Traditionally, PGA has been used to define the strength of ground shaking, but the correlation between PGA and ED is low. Because PGA is a measure of high frequencies, it is suitable only for non-engineered brittle structures. For engineered structures, PGV or ED is better suited to define the strength of ground shaking. In past earthquakes, "pulse-like" ground motions with high PGV have been found to be very destructive (Heaton et al. 1995, Hall 1998, Malhotra 1999, Gupta and Krawinkler 2000, Olsen et al. 2008). It is not the "pulse" but the resulting PGV that is the cause of high ED, hence high damage potential of a ground motion.

It is proposed that ground motions with $1 \le ED < 3\,(\mathrm{m/s})^2$ be considered strong, those with $3 \le ED < 10\,(\mathrm{m/s})^2$ be considered very strong, and those with $ED \ge 10\,(\mathrm{m/s})^2$ be considered extremely strong. Per Equation (1-16), ED of 1, 3, and 10 $(\mathrm{m/s})^2$ correspond to PGV of 35, 61, and 112 cm/s, respectively. Table 1-1 lists

Table 1-1. Proposed Definition of Strength of Ground Motion for Engineered Structures.

Weak	$ED < 1\,(\mathrm{m/s})^2$	$PGV < 35\,\mathrm{cm/s}$
Strong	$1 \le ED < 3\,(\mathrm{m/s})^2$	$35 \le PGV < 61\,\mathrm{cm/s}$
Very strong	$3 \le ED < 10\,(\mathrm{m/s})^2$	$61 \le PGV < 112\,\mathrm{cm/s}$
Extremely strong	$ED \ge 10\,(\mathrm{m/s})^2$	$PGV \ge 112\,\mathrm{cm/s}$

Table 1-2. Strongest Recorded Ground Motions.

No.	Earthquake	Recording station	R_{jb} (km)	ED $(m/s)^2$	T_{cg} (s)	PGV_n
1	1994 **M** 6.7 Northridge, CA	Tarzana Cedar Hill	0.37	10.3	0.57	0.39
2	1989 **M** 6.9 Loma Prieta, CA	LGPC	0	11.8	1.64	0.42
3	1999 **M** 7.6 Chi-Chi, Taiwan	TCU065	0.59	14.9	2.72	0.48
4	1994 **M** 6.7 Northridge, CA	Sylmar Converter	0	15.4	1.54	0.57
5	1999 **M** 7.6 Chi-Chi, Taiwan	CHY080	0.11	15.9	1.01	0.54
6	1995 **M** 6.8 Kobe, Japan	Takatori	1.46	20.0	1.44	0.92
7	1999 **M** 7.6 Chi-Chi, Taiwan	TCU084	0	20.0	1.28	0.59
8	2023 **M** 7.8 Nurdagi, Türkiye	TK3123	2.4	20.1	2.37	0.76
9	1999 **M** 7.6 Chi-Chi, Taiwan	TCU052	0	21.4	3.95	0.53
10	2004 **M** 6.6 Niigata, Japan	NIG019	0	22.5	0.74	0.58
11	1999 **M** 7.6 Chi-Chi, Taiwan	TCU068	0	36.2	6.48	0.45

the proposed definition of the strength of ground motion for engineered structures. Table 1-2 lists the 11 strongest ground motions recorded throughout the world up to this point, based on available strong-motion data (PEER 2022, COSMOS 2022, CESMD 2022, AFAD 2023) and literature review (e.g., Strasser and Bommer 2009). In Table 1-2, R_{jb} refers to minimum distance from surface projection of the fault, aka the Joyner–Boore distance. By rough estimates, 5,000 instruments, with an average recording time of 40 years, have provided $5{,}000 \times 40 = 200{,}000$ instrument-year of recorded data. Most of these instruments are in populated, high-seismic areas with relatively higher probability of ground shaking. Hence, there is negligible chance of exceeding these ground motions at specific sites anywhere in the world. A 1/200,000 chance of exceedance per year is negligible compared to the 1/5,000 chance of collapse per year accepted by current design standards (Luco et al. 2007). It is probable that future earthquakes will produce even stronger ground motions, but those will be from a longer dataset of > 200,000 instrument-year. Therefore, the probability of exceeding the strongest ground motions in Table 1-2 will still remain negligible.

1.11 Cyclic Demand

During an earthquake, a structure moves back and forth in a cyclic manner. In other words, the structure experiences half cycles of various amplitudes. The peak-deformation half cycle causes most damage to the structure, but smaller amplitude half cycles collectively can also cause significant damage to the structure (Miner 1945, Krawinkler et al. 1983, Malhotra 2002). Based on experiments, it is generally assumed that the damage caused by each half cycle is proportional to the square of its amplitude. Using this assumption, it is possible to calculate the equivalent number of uniform-amplitude cycles experienced by the structure (Malhotra 2002), as described next.

Referring to Figure 1-10, a 5% damped structure of period 1 s experiences maximum response in a direction that is $\phi = 70°$ clockwise from north (N70E). From the deformation histories in east and north directions, shown in Figure 1-9, the deformation history in the maximum-horizontal direction is calculated as follows:

$$u_{\max} = u_e \sin\phi + u_n \cos\phi \tag{1-17}$$

Note that the direction of maximum-horizontal response ϕ is not fixed. It changes with ground motion, and it also changes with period and damping. Figure 1-20 shows the deformation history of 5% damped, 1 s period structure in the maximum-horizontal direction.

Each half-cycle in Figure 1-20 is assigned a weight proportional to the square of its amplitude. Half-cycle with the peak amplitude of 9.96 in. is assigned a weight of 1. Half-cycle with an amplitude of 6.54 in. is assigned a weight of $(6.54/9.96)^2 = 0.43$, and so on. Weights of all half-cycles are added and divided by 2 to obtain the equivalent number of uniform-amplitude "full" cycles. A 5% damped, 1 s period structure experiences 1.7 cycles of peak amplitude 9.96 in. The number of cycles is similarly computed for other values of period and damping. Figure 1-21 shows 5%, 10%, and 20% damped cyclic-demand response spectra (CDRS) of TAPS ground motion in maximum-horizontal direction. Note that the number of cycles reduces with increase in damping, and it also reduces with increase in period. It will become clearer in later chapters that ductile and hysteretic structures can significantly increase their period and damping during strong shaking. That is why such

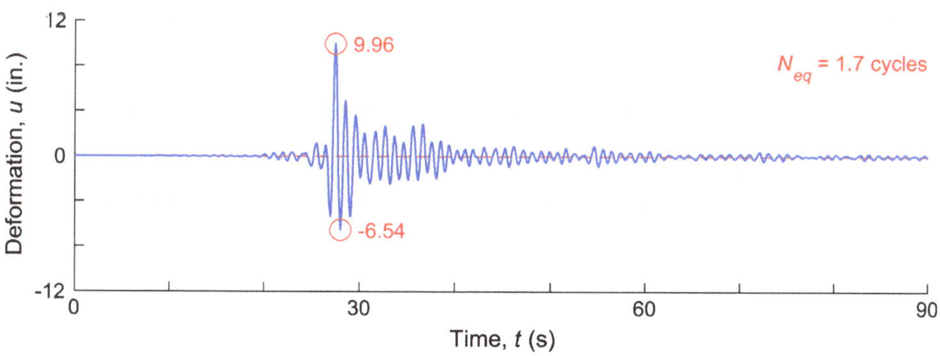

Figure 1-20. Deformations of the single-degree-of-freedom system of period $T = 1$ s, and damping $\zeta = 0.05$ (5% of critical), in the maximum-horizontal (RotD100) direction, subjected to TAPS ground motion.

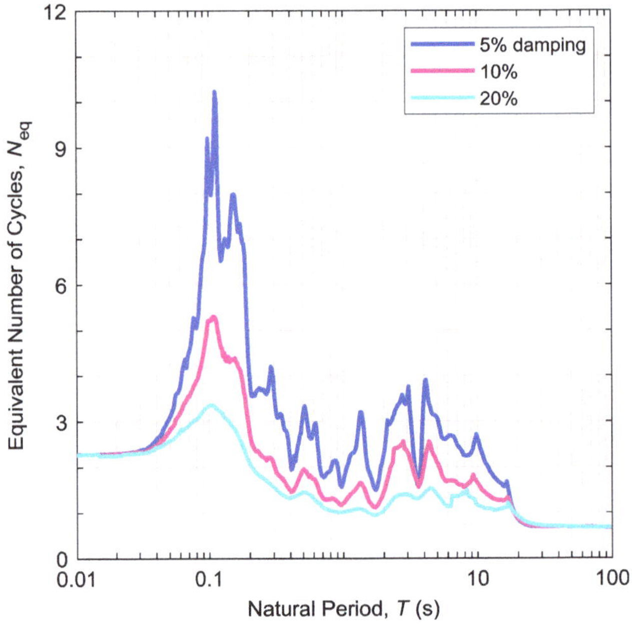

Figure 1-21. Cyclic-demand response spectra of TAPS ground motion shown in Figures 1-1 to 1-6 for various values of damping.

structures experience fewer cycles than brittle and nonhysteretic structures. The definition of ductile and hysteretic structures will become clear in Chapter 3.

Cyclic demand is typically ignored in seismic design, but it can be significant for large magnitude earthquakes. A practical way of considering cyclic demand is discussed in this book.

1.12 Indirect Parameters of Ground Motion

ADRS and CDRS directly affect the response of structures. There are many other parameters that indirectly affect the response of structures. For example,

- *PGA* is an indirect measure of short-period portion of ADRS.
- Significant duration (Trifunac 1975) is an indirect measure of cyclic demand.
- Arias Intensity (Arias 1970) is an indirect measure of *PGA* and duration combined.
- Directivity pulses (Malhotra 1999) are indirect measures of *PGV*, which correlates with *ED*.
- Cumulative absolute velocity (EPRI 1991) is an indirect measure of *PGA* and duration, like Arias Intensity.

As far as possible, only direct parameters of ground motion should be used in design and evaluation of structures. In subsequent chapters, only direct parameters of ground motion are used.

1.13 Summary

1. The response spectrum of a ground motion provides information that can be directly used by an engineer.
2. The tripartite plot is a very concise and elegant representation of the response spectrum.
3. The acceleration–deformation plot of response spectrum (ADRS) is helpful in predicting the response of nonlinear systems.
4. Deformation of a nonlinear system reduces with (1) increase in normalized strength (lateral strength/mass) and (2) increase in damping.
5. The cyclic-demand spectrum is helpful in estimating the equivalent number of cycles of peak response.
6. The equivalent number of cycles generally reduces with (1) increase in period and (2) increase in damping.
7. Energy demand, *ED*, defines the size of the response spectrum; energy centroid T_{cg} and normalized velocity PGV_n define its shape.
8. *ED* correlates well with the square of peak ground velocity PGV. *ED* does not correlate well with the peak ground acceleration PGA.
9. It is proposed that ground motions with $ED > 1\ (\mathrm{m/s})^2$ be considered strong, those with $ED > 3\ (\mathrm{m/s})^2$ be considered very strong, and those with $ED > 10\ (\mathrm{m/s})^2$ be considered extremely strong.
10. Only a handful of ground motions with $ED > 10\ (\mathrm{m/s})^2$ have been recorded so far.
11. The important engineering parameters of a ground motion are its amplitude and cyclic-demand spectra (ADRS and CDRS) for various values of damping.
12. Strongest ground motions listed in Table 1-2 have negligible chance of exceedance at specific sites anywhere in the world.

References

Arias, A. 1970. “A measure of earthquake intensity.” In *Proc., Conf. on Seismic Design of Nuclear Power Plants*. Boston: Massachusetts Institute of Technology, 438–483.

Baker, J. W., and C. A. Cornell. 2006. “Correlation of response spectral values for multi-component ground motions.” *Bull. Seismol. Soc. Am.* 96 (1): 215–227.

Boore, D. M., J. Watson-Lamprey, and N. A. Abrahamson. 2006. “Orientation-independent measures of ground motion.” *Bull. Seismol. Soc. Am.* 96 (4A): 1502–1511.

CESMD (Center for Engineering Strong Motion Data). 2022. Center for Engineering Strong Motion Data. Cooperative center by USGS and CGS. Accessed September 10, 2023, https://www.strongmotioncenter.org/.

Chopra, A. K. 2017. *Dynamics of structures.* 5th ed. London: Pearson Publications.

Clough, R. W., and J. Penzien. 2003. *Dynamics of structures.* 3rd ed. Berkeley, CA: Computers and Structures, Inc.

COSMOS. 2022. Strong-Motion Virtual Data Center (VDC). Accessed November 16, 2022. https://www.strongmotioncenter.org/vdc/scripts/plot.plx?stn=3037&evt=814.

EPRI (Electrical Power Research Institute). 1991. "*Standardization of the cumulative absolute velocity.* EPRI TR-100082-T2. Palo Alto, CA: EPRI.

Gupta, A., and H. Krawinkler. 2000. "Behavior of ductile SMRFs at various seismic hazard levels." *J. Struct. Eng.* 126 (1): 98–107.

Hall, J. F. 1997. "Seismic response of steel frame buildings to near-source ground motions." *Tech. Rept. EERL 97-05,* Pasadena, CA: California Institute of Technology.

Heaton, T. H., J. F. Hall, D. J. Wald, and M. W. Halling. 1995. "Response of high-rise and base-isolated buildings to a hypothetical M_w 7.0 blind thrust earthquake." *Science* 267 (S195): 206–211.

Hudson, D. E. 1979. *Reading and interpreting strong motion accelerograms.* Richmond, CA: Earthquake Engineering Research Institute.

Krawinkler, H., M. Zohrei, B. Lashkari-Irvani, N. G. Cofie, et al. 1983. *Recommendation for experimental studies on the seismic behavior of steel components and materials.* NSF/CEE-83220. Stanford, CA: Stanford University.

Luco, N., B. R. Ellingwood, R. O. Hamburger, J. D. Hooper, et al. 2007. "Risk-targeted vs. current seismic design maps for the conterminous United States." In *Proc., SEAOC 76th Ann. Conv.* Sacramento, CA: Structural Engineers Association of California.

Malhotra, P. K. 1999. "Response of buildings to near-field pulse-like ground motions." *Earthq. Eng. Struct. Dyn.* 28 (11): 1309–1326.

Malhotra, P. K. 2001. "Response spectrum of incompatible acceleration, velocity and displacement histories." *Earthq. Eng. Struct. Dyn.* 30 (2): 279–286.

Malhotra, P. K. 2002. "Cyclic-demand spectrum." *Earthq. Eng. Struct. Dyn.* 31 (7): 1441–1457.

Malhotra, P. K. 2021. "Ground motions from past earthquakes." Chap. 1 in *Seismic analysis of structures and equipment.* Cham, Switzerland: Springer. https://doi.org/10.1007/978-3-030-57858-9_1.

Malhotra, P. K. 2022. *Site-specific ground motions for seismic design of buildings and other structures.* Reston, VA: ASCE.

Miner, M. A. 1945. "Cumulative damage in fatigue." *J. Appl. Mech.* 12 (A): 159A–164 (1945).

Newmark, N. M., and W. Hall. 1982. *Earthquake spectra and design.* Richmond, CA: Earthquake Engineering Research Institute.

Olsen, A. H., B. T. Aagaard, and T. H. Heaton. 2008. "Long-period building response to earthquakes in the San Francisco Bay Area." *Bull. Seismological Society of America* 98 (2): 1047–1065.

PEER (Pacific Earthquake Engineering Research Center). 2022. "PEER ground motion database." Berkeley, CA: University of California, PEER. Accessed September 10, 2023. https://ngawest2.berkeley.edu/.

Strasser, F. O., and J. J. Bommer. 2009. "Large-amplitude ground-motion recordings and their interpretations." *Soil Dyn, Earthq. Eng.* 29 (10): 1305–1329.

Trifunac, M. D., and A. G. Brady. 1975. "A study of the duration of strong earthquake ground motion." *Bull. Seismol. Soc. Am.* 65 (3): 581–626.

CHAPTER 2

Comparison with Predicted Ground Motions

2.1 Introduction

Future ground motions are predicted by a probabilistic seismic hazard analysis (PSHA) (Esteva 1970, Cornell 1971, Reiter 1991, McGuire 2004). In the United States, PSHA is performed by the US Geological Survey (USGS) (Petersen et al. 2019). Per PSHA, any level of ground motion can be exceeded in the future. Lower level of ground motion is exceeded at a higher rate [shorter mean return period (MRP)], and higher level of ground motion is exceeded at a lower rate (longer MRP). The probabilistic seismic hazard analysis is highly subjective because there are diverse opinions regarding geoscientific information such as (1) the magnitude, location, and occurrence of future earthquakes; and (2) ground motions produced by specific earthquakes. To obtain stable results, diverse opinions are aggregated by following a complex procedure outlined in Senior Seismic Hazard Analysis Committee (SSHAC) guidelines (Budnitz et al. 1997, Hanks et al. 2009). In this chapter, the predicted ground motions are compared with the strongest recorded ground motions.

2.2 San Francisco, California, Ground Motion

A site with latitude 37.7749° N and longitude 122.4194° W is selected in the San Francisco Bay area. Per USGS (2023), the ground shaking hazard at the site is controlled by a magnitude **M** 7.9 earthquake on San Andreas fault at 13 km from the site. The site is assumed to be on Class D soil as defined by ASCE 7-22. A 10,000-year MRP response spectrum is generated per the 2018 USGS analysis (Shumway et al. 2021). The US Geological Survey provides peak ground acceleration PGA and peak pseudo-acceleration PPA from 0.01 to 10 s period. Recall from Chapter 1 that $PGA = PPA(0\,\text{s})$. Therefore, the PGA and PPA provided by USGS can be used to construct the response spectrum from 0 to 10 s period. PGA and PPA provided by USGS are rotated-median (RotD50) of horizontal ground motion. They are converted to maximum horizontal (RotD100) by using conversion factors in Figure 1-19 of Malhotra (2022). To extend the response spectrum beyond 10 s, a smooth response spectrum, described in Section 1.12 of Malhotra (2021), is generated

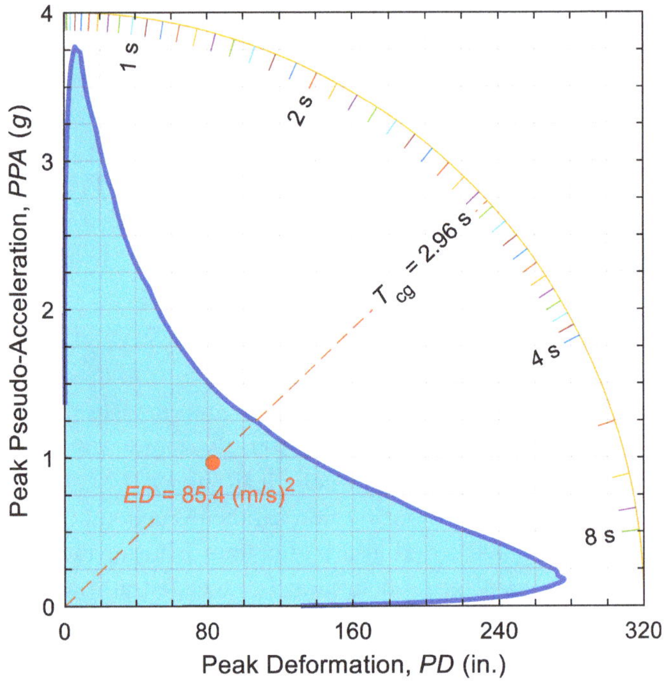

Figure 2-1. Five percent damping, 10,000-year MRP acceleration–deformation response spectrum for a site in the San Francisco Bay area.

from assumed values of PGA, PGV, and PGD such that it provides a smooth extension beyond 10 s. Alternatively, the response spectrum could be extended by adding a displacement-sensitive region as described in Section 2.6 of Malhotra (2022).

Figure 2-1 shows a plot of the 5% damping, 10,000-year MRP ADRS (acceleration–deformation response spectrum) per the USGS analysis (Shumway et al. 2021). The peak values of ground acceleration, velocity, and displacement, inferred from the response spectrum, are $PGA = 1.35g$, $PGV = 100$ in./s, and $PGD = 121$ in. The energy demand $ED = 85.4\,(\text{m/s})^2$, energy centroid $T_{cg} = 2.96$ s, and normalized velocity $PGV_n = 0.4$. The 10,000-year MRP predicted ground motion is extremely strong, low frequency, and broad banded. Recall from the Preface and Chapter 1 that 200,000 instrument-year of recorded data have been collected so far. A 10,000-year MRP ground motion has nearly 100% chance of being exceeded in a 200,000-year ground motion data, but a ground motion with $ED \geq 85.4\,(\text{m/s})^2$ has never been recorded anywhere in the world.

2.3 Hickman, Arkansas, Ground Motion

Next, a site with latitude 35.9414° N and longitude 89.7211° W is selected in Hickman, Arkansas. Per USGS (2023), the ground shaking hazard at the site is controlled by a magnitude **M** 7.5 earthquake in the New Madrid Seismic Zone at 15 km from the site. The site is assumed to be on Class D soil as defined by ASCE 7-22. A 10,000-year MRP response spectrum is generated per the 2018 USGS analysis (Shumway et al. 2021). Figure 2-2 shows a plot of the 5% damping, 10,000-year MRP ADRS per USGS. The peak values of ground acceleration, velocity, and displacement, inferred from the response spectrum, are $PGA = 2.35g$, $PGV = 98.7$ in./s, and $PGD = 87.6$ in. With $ED = 70.4\ (\text{m/s})^2$, $T_{cg} = 2.07$ s, and $PGV_n = 0.35$, the predicted ground motion is extremely strong, has low to medium frequency, and is broad banded. A ground motion with such high energy demand has never been recorded anywhere in the world.

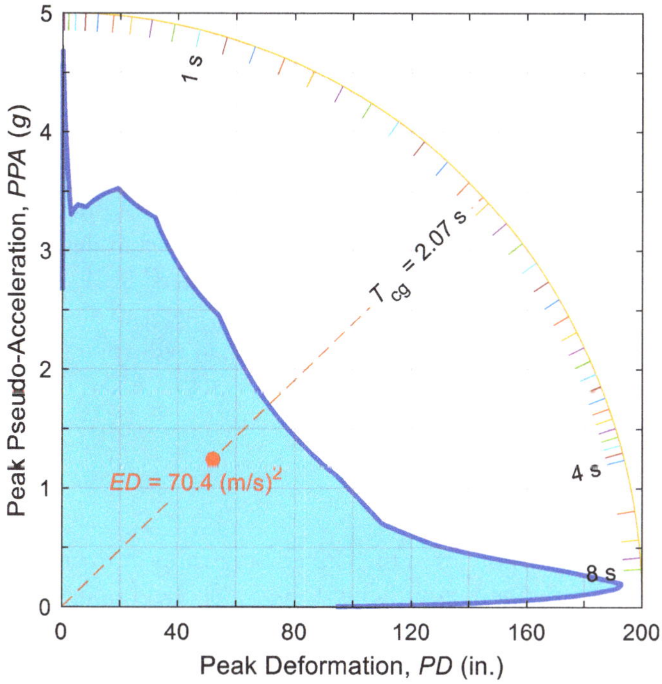

Figure 2-2. Five-percent damping, 10,000-year MRP acceleration–deformation response spectrum for a site in Hickman, Arkansas.

2.4 Discussion

The 10,000-year MRP predicted ground motions at two sites in the United States are much stronger than any ground motions in 200,000 instrument-year of recorded data. Note that a 10,000-year MRP ground motion has nearly 100% chance of exceedance in a 200,000-year ground motion data set. This suggests that the response spectra of extreme ground motions are overpredicted by the PSHA. The *ED* of 20,000-year MRP ground motions is completely unrealistic, even though a 20,000-year ground motion also has nearly 100% chance of exceedance in a 200,000-year of recorded data.

Overprediction of hazard (hence risk) is not in the best interest of public for the following reasons:

1. Overprediction of one hazard can take resources away from other hazards. The resources are always limited, and they should be allocated to minimize the overall risk. If one hazard is more overpredicted than others, it will result in inefficient allocation of resources.
2. Overprediction of hazard increases the cost of transferring (insuring) the risk.
3. Overprediction of hazard can delay or cancel a project, thus depriving the public of its benefits.
4. Overprediction of hazard causes others to doubt the results of hazard analysis and to not take them seriously. For example, in selecting design ground motions, engineers use various arbitrary factors to reduce the PSHA predictions (Malhotra 2005, 2022). Some of these factors are (1) risk factors and deterministic limit to obtain maximum considered earthquake risk-targeted (MCE_R) ground motions from probabilistic ground motions, (2) two-thirds factor to obtain design ground motions from MCE_R ground motions, and (3) load reduction factor R of up to 8 to reduce the design ground motions. These factors make the seismic design quite arbitrary.

2.5 Summary

1. The extreme ground motions in the United States appear to be overpredicted by PSHA. That is partly the reason behind the widely held opinion that structures cannot be designed to withstand the strongest ground motions.
2. Overprediction of hazard (and risk) is harmful because it (1) increases the cost of insurance, (2) results in inefficient allocation of resources, (3) deprives public of the benefits of projects that are deemed “too risky” or “too costly” owing to overprediction, and (4) causes others to doubt the predictions and to not take them seriously.

References

ASCE. 2022. *Minimum design loads for building and other structures.* ASCE-7-22. Reston, VA: ASCE.

Budnitz, R. J., G. Apostolakis, D. M. Boore, L. S. Cluff, et al. 1997. "Recommendations for probabilistic seismic hazard analysis: Guidance on uncertainty and use of experts." *NUREG/CR-6372,* Vol. 1. Palo Alto, CA: Electric Power Research Institute.

Cornell, C. A. 1971. "Probabilistic analysis of damage to structures under seismic loads." In *Dynamic waves in civil engineering,* edited by D. A. Howells, I. P. Haigh, and C. Taylor. New York: Wiley, 473–488.

Esteva, L. 1970. "Seismic risk and seismic design decisions, in seismic design of nuclear power plants." In *Seismic design of nuclear power plants,* edited by R. J. Hansen. Cambridge, MA: MIT Press.

Frankel, A., R. Chen, M. Petersen, M. Moschetti, et al. 2015. "2014 update of the Pacific Northwest portion of the U.S. national seismic hazard maps. *Earthquake Spectra* 31 (1): suppl. 2015.

Hanks, T. C., N. A. Abrahamson, D. M. Boore, K. J. Coppersmith, et al. 2009. "Implementation of the SSHAC guidelines for Level 3 and 4 PSHAs—Experience gained from actual applications." *Open-File Report 2009-1093.* Reston, VA: US Geological Survey.

Malhotra, P. K. 2005. "Return period of design ground motions." *Seismol. Res. Lett.* 76 (6): 693–699.

Malhotra, P. K. 2022. *Site-specific ground motions for seismic design of buildings and other structures.* Reston, VA: ASCE.

McGuire, R. K. 2004. *Seismic hazard and risk analysis.* Oakland, CA: Earthquake Engineering Research Institute.

Petersen, M. D., A. M. Shumway, P. M. Powers, C. S. Mueller, et al. 2019. "The 2018 update of the US National Seismic Hazard Model: Overview of model and implications:" *Earthquake Spectra* 36 (1): 5–41. https://doi.org/10.1177/8755293019878199.

Reiter, L. 1991. *Earthquake hazard analysis—issues and insights.* New York: Columbia University Press.

Shumway, A. M., B. S. Clayton, and K. S. Rukstales. 2021. "Data release for additional period and site class data for the 2018 national seismic hazard model for the conterminous United States (version 1.2, May 2021)." US Geological Survey data release. https://doi.org/10.5066/P9RQMREV.

USGS (US Geological Survey). 2023. *Unified hazard tool.* Earthquake hazards program. Reston, VA: USGS. Accessed September 10, 2023. https://earthquake.usgs.gov/hazards/interactive/.

CHAPTER 3

Seismic Characteristics of Buildings

3.1 Introduction

Strength, deformability, and damping play equally important roles in seismic design (Malhotra 2021). Buildings are not expected to withstand extreme ground motions by strength alone. Two example buildings are selected: (1) 3-story steel special moment frame (SMF) building and (2) 20-story steel eccentrically braced frame (EBF) building. Both buildings are designed to possess high strength, deformability, and damping by utilizing ordinary materials and established technologies. The amounts of structural material used are not significantly different from those used in modern engineered buildings in high-seismic areas such as California.

3.2 Three-Story Steel Special Moment Frame Building

Figure 3-1 shows the sketch of a 3-story steel SMF building. A similar building has been used as a design example in SEAOC (2012) manual. The building has two 3-bay steel SMF in each orthogonal direction. Special moment frames are tied together at each level by concrete-on-steel-deck floor slabs that act as "rigid" diaphragms. By definition of SMF (AISC 2016a, b; Hamburger et al. 2009), (1) beam-column connections are strong enough to resist the moments and shears from fully yielded and strain-hardened beams; and (2) columns are strong enough to remain elastic, except at the base, even after the beams have fully yielded and strain-hardened. For this building, column bases are designed to have pin connections with the foundation. The lumped weights at second, third, and fourth (roof) levels are 1,320, 1,320, and 660 kip, respectively. The weights include dead load and expected live load. The weight of the entire building is $W = 3{,}300$ kip. Each story is 12 ft high, and each bay is 30 ft long. The total height of the building is $H = 432$ in., and the height of its center of gravity (CG) is $H_{CG} = 259$ in. The cross section of two outer columns is W30 × 173 and that of two inner columns is W33 × 291. The cross section of all beams is W30 × 116. Structural steel is ASTM A992 with minimum yield stress of $F_y = 50$ ksi and expected yield-stress of $F_{ye} = 55$ ksi. The first-mode period of the building from linear-elastic analysis (CSI 2011) is $T_1 = 0.777$ s. This is significantly longer than the period from approximate formula in ASCE 7-22 or from measurements during

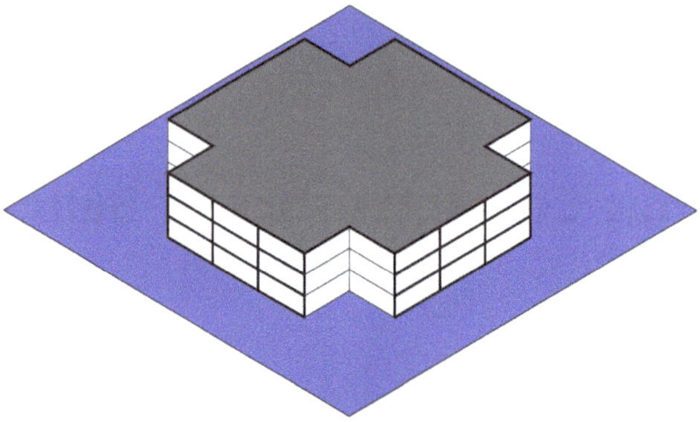

Figure 3-1. Three-story steel special moment frame building.

ambient wind because those include contributions from nonstructural components such as infills and claddings, as well as from the gravity load resisting system. In this book, only the contributions of designated lateral load resisting system to the strength and damping of the building are considered, because contributions from other sources are unreliable.

The expected plastic moment capacity M_p of beam cross section is

$$M_p = Z_p \cdot F_{ye} \tag{3-1}$$

where Z_p is the plastic modulus of beam cross section. Substituting $Z_p = 378\,\text{in.}^3$ for W30 × 116, and $F_{ye} = 55\,\text{ksi}$ for ASTM A992 steel into Equation (3-1), $M_p = 20{,}790\,\text{kip-in.}$ is obtained.

3.3 Pushover Curve for 3-Story Building

The lateral strength and deformability of a building are defined by its pushover curve, which is a plot between the lateral force and the deflection of building's CG (center of gravity). As the building is pushed laterally, bending moments rise in the beams. When the moments at the ends of beams reach the plastic moment capacities of beam cross sections, plastic hinges appear at the ends of beams. With further deflection of the CG, the beams continue to yield, but the connections remain intact, and the columns remain elastic. Yielding occurs only in the beams, and it is flexural in nature. For the sake of simplicity, it is assumed that plastic hinges form at the centerlines of columns. In other words, "rigid" offsets at the ends of beams are ignored. Rigid offsets increase the lateral strength of SMF but reduce its deformability (Malhotra 2021). Therefore, the net effect of modeling rigid offsets is not significant in most cases.

Figure 3-2 shows the plastic moments in a fully yielded SMF. Disregarding the effect of gravity, the lateral force at the CG of a fully yielded SMF is

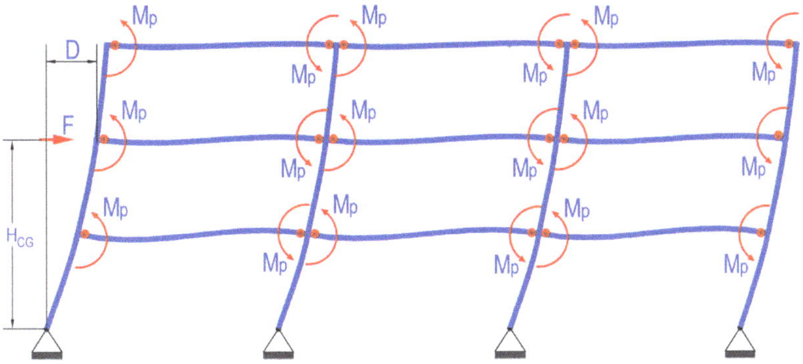

Figure 3-2. Plastic moments in a fully yielded special moment frame.

$$F_p = \frac{\sum M_p}{H_{CG}} \tag{3-2}$$

in which $\sum M_p$ is the summation of plastic moments in all SMF connections. For this building, there are 12 SMF connections (six per frame) at each level. There is a total of 36 SMF connections in each orthogonal direction, and M_p is the same for all beams. Upon substituting $\sum M_p = 36 \times 20{,}790$ kip-in., and $H_{CG} = 259$ in. into Equation (3-2), $F_p = 2{,}890$ kip is obtained.

The elastic stiffness, before the formation of any plastic hinge, is given by the following expression:

$$k = M\left(\frac{2\pi}{T_1}\right)^2 = \frac{W}{g}\left(\frac{2\pi}{T_1}\right)^2 \tag{3-3}$$

where M = seismic mass = W/g. Substituting $T_1 = 0.777$ s = first-mode period of the building, $W = 3{,}300$ kip, and $g = 386$ in./s^2 into Equation (3-3), $k = 559$ kip/in. is obtained. The maximum elastic deflection is

$$D_e = \frac{F_p}{k} \tag{3-4}$$

Substituting $F_p = 2{,}890$ kip and $k = 559$ kip/in. into Equation (3-4), $D_e = 5.17$ in. is obtained. The elastic portion of the pushover curve ($D \leq D_e$) is generated by using the following equation:

$$F = k \cdot D \tag{3-5}$$

For $D \leq D_e$, the force increases linearly with D, but the plastic rotation in beams θ_p^b remains zero. For $D > D_e$, the force remains fixed at $F = F_p$, but θ_p^b increases with increase in D. Because plastic hinges are assumed at the centerlines of columns, the plastic drift ratio $\theta_p = \theta_p^b$. The total deflection D is

$$D = D_e + \theta_p \cdot H_{CG} \tag{3-6}$$

The limiting value of $\theta_p = \theta_p^b$ depends on the length of beam L, and compactness of the beam cross section defined by (1) width/thickness of flange, and (2) depth/thickness of web. Per Table 9-7.1 of ASCE 41-17 (ASCE 2017), the limiting value is $\theta_p = \theta_p^b = 0.0844$ radian. A collapse is assumed to occur when plastic drift ratio reaches $\theta_p = 0.0844$ radian. Therefore, per Equation (3-6), the ultimate deflection of the CG at collapse is $D_{ult} = 5.17 + 0.0844 H_{CG} = 27$ in. The ultimate drift ratio (elastic plus plastic) is $\theta_{ult} = D_{ult}/H_{CG} = 0.104$ radian. Figure 3-3 shows the simplified pushover curve for the building from Equations (3-2) to (3-6). The ductility ratio μ is the ratio between the ultimate deflection and the elastic deflection. For this building, $\mu = D_{ult}/D_e = 27/5.17 = 5.23$. The ductility ratio, μ, determines the damping of a hysteretic structure as seen later.

The pushover curve of Figure 3-3 has an ideal (elastic-fully-plastic) shape. It assumes that the building transitions from elastic to fully plastic state abruptly. In reality, the transition from elastic to fully plastic state is gradual. It can be shown through analysis that the building starts yielding at a lateral force of 1,240 kip (and a lateral deflection of 2.3 in.) when the flanges of first-story beams in the outer bays start yielding. With increase in deflection, the yielding spreads to the flanges of other beams in the building, and then to the webs of various beams until the building reaches a fully yielded state. Figure 3-4 compares the ideal pushover curve with the

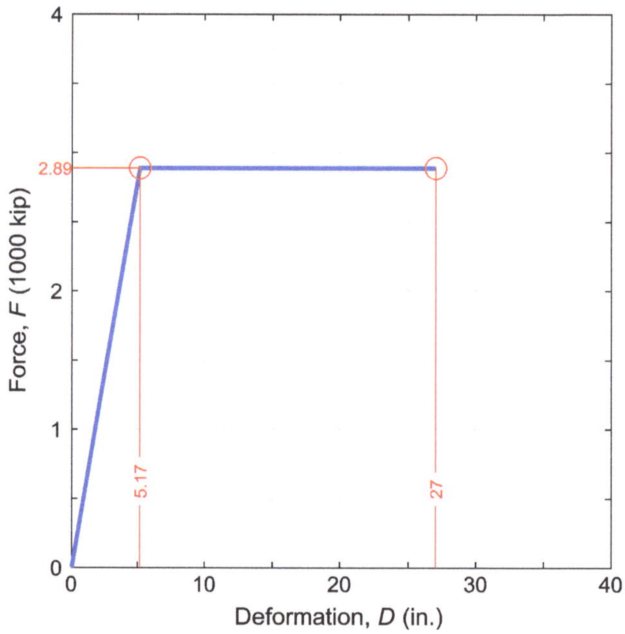

Figure 3-3. Simplified pushover curve for the 3-story steel special moment frame building shown in Figure 3-1.

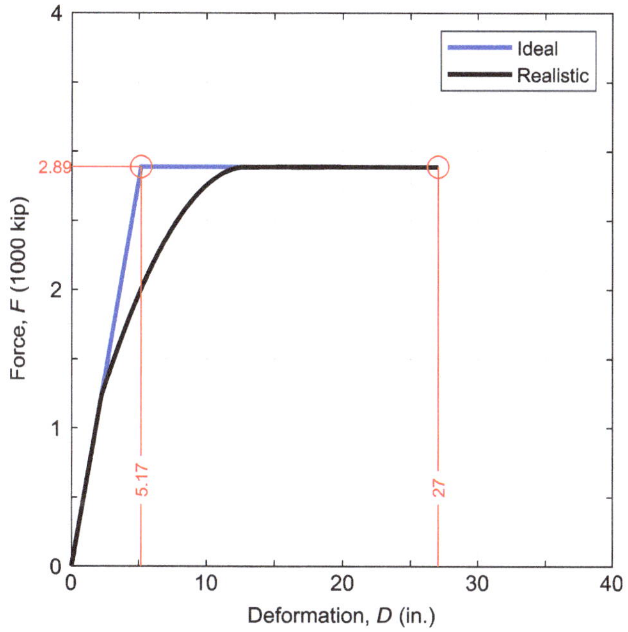

Figure 3-4. Ideal versus realistic pushover curves for the 3-story steel special moment frame building.

realistic pushover curve. The gradual yielding of the building has some effect on its damping as seen later. Other than that, the use of ideal pushover curve does not introduce significant approximation in the analysis.

The pushover curve of Figure 3-3 ignores any increase in plastic moment with increase in plastic rotation due to strain hardening. The pushover curve of Figure 3-3 also ignores the effect of gravity, aka the P-Δ effect (Paulay 1978). The pushover curve is refined as follows:

Strain hardening. Due to strain hardening, the plastic moment M_p increases with increase in plastic rotation θ_p^b. For steel SMF, AISC 341-16 (AISC 2016b) suggests a 10% increase in strength due to strain hardening. Figure 3-5 shows the pushover curve with strain hardening. Strain hardening causes the ultimate lateral strength of the building to increase from 2,890 to 3,180 kip. The fully yielded, strain-hardened moment capacity of beams is $20{,}790 \times 1.1 = 22{,}870$ kip-in. Columns and connections should be designed to resist this moment and the corresponding shear.

P-Δ effect. Referring to Figure 3-2, notice that the weight of the deflected building applies an additional moment that is in the same direction as the moment applied by the lateral force F. Both moments are resisted by the SMF. Hence, the weight of the structure reduces the lateral force F for a given deflection D. The P-Δ effect is considered as follows: (1) For a given value of D, F is read from Figure 3-5; and (2) F is reduced by an amount DW/H_{CG}, where W is the weight of the building.

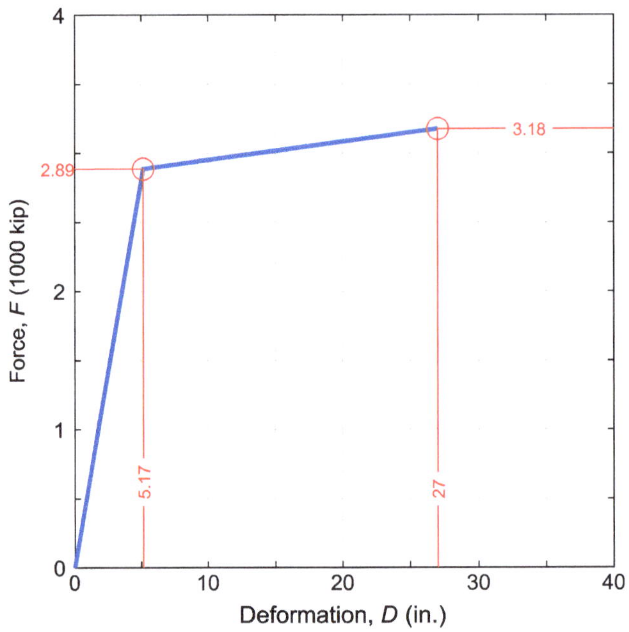

Figure 3-5. Pushover curve for the 3-story steel special moment frame building with strain-hardening.

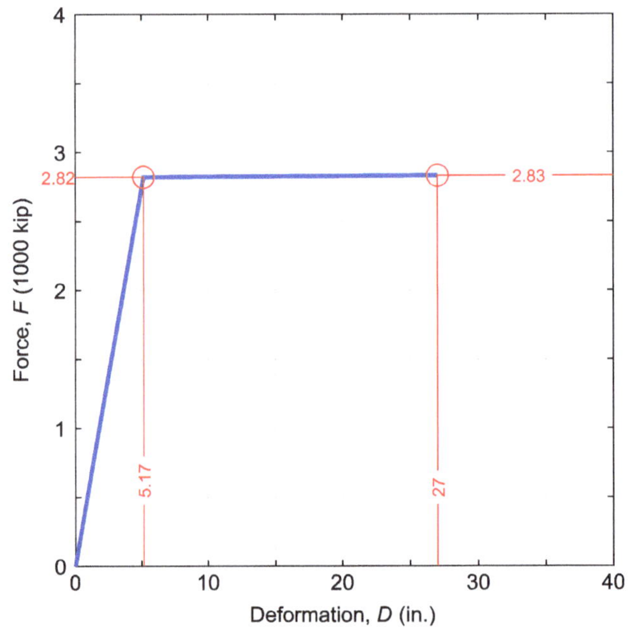

Figure 3-6. Pushover curve for the 3-story steel special moment frame building with strain hardening and P-Δ effect.

Figure 3-6 shows the refined pushover curve after considering strain-hardening and P-Δ. Due to P-Δ, the yield strength of the building reduces from 2,890 to 2,820 kip, and the ultimate strength of the building reduces from 3,180 to 2,830 kip. For this building, the ultimate strength of 2,830 kip is only 0.35% higher than the yield strength of 2,820 kip. It is important for the pushover curve to have a positive slope (i.e., the lateral force should increase with increase in deflection). If the pushover curve has a zero or negative slope, even a slight torsional eccentricity will cause SMF on one side of the building to yield and the SMF on the opposite side of the building to remain elastic. This would cause the building to have a large torsional response, thereby reducing its deformation capacity (Malhotra 2021).

3.4 Capacity Curve for 3-Story Building

The pushover curve of Figure 3-6 is converted to the capacity curve by dividing the force along the vertical axis by the total seismic mass M = 3,300 kip/g. Figure 3-7 shows a plot of the capacity curve. The vertical axis in Figure 3-7 is pseudo-acceleration PA in g. Recall from Chapter 1 that the natural period is related to the ratio between the deformation and pseudo-acceleration, as follows:

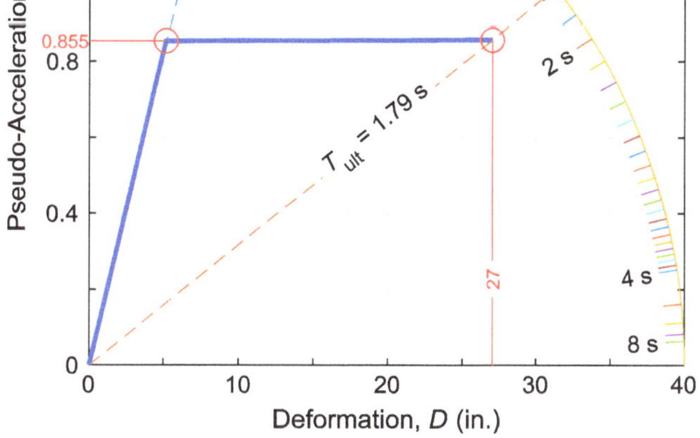

Figure 3-7. Capacity curve for the 3-story steel special moment frame building shown in Figure 3-1.

$$T = 2\pi\sqrt{\frac{D}{PA}} \tag{3-7}$$

With the help of Equation (3-7), radial tick-marks corresponding to various periods are drawn in Figure 3-7. Notice that the period of the building at onset of yielding, at 5.17 in. deflection, is 0.79 s. This is slightly longer than the period of 0.777 s calculated from the linear-elastic analysis. The slight increase is due to P-Δ effect. After yielding, the period increases with increase in deflection. The capacity curve ends at ultimate deflection of $D_{ult} = 27$ in. when the plastic rotation in beams reaches its limiting value of $\theta_p^b = 0.0844$ radian and the building collapses. When the CG of the building deflects 27 in., the roof deflection is about $27/H_{CG} \times H = 27/259 \times 432 = 45$ in. Therefore, 45 in. clearance is needed all around the building for it to have a deformability of 27 in.

The effective period of the building at $D = D_{ult}$ is $T_{ult} = 1.79$ s. The period of the 3-story SMF building is not fixed during ground shaking; it ranges from 0.79 to 1.79 s. The area below the capacity curve (Figure 3-8) is the seismic toughness ST (Malhotra 2021). Seismic toughness ST has the units of energy/mass. For the 3-story steel SMF building, $ST = 5.22\ (\text{m/s})^2$. In general, structures with higher ST perform better during earthquakes than those with lower ST. An approximate value of ST can be obtained by adding the toughness of all structural elements. There are

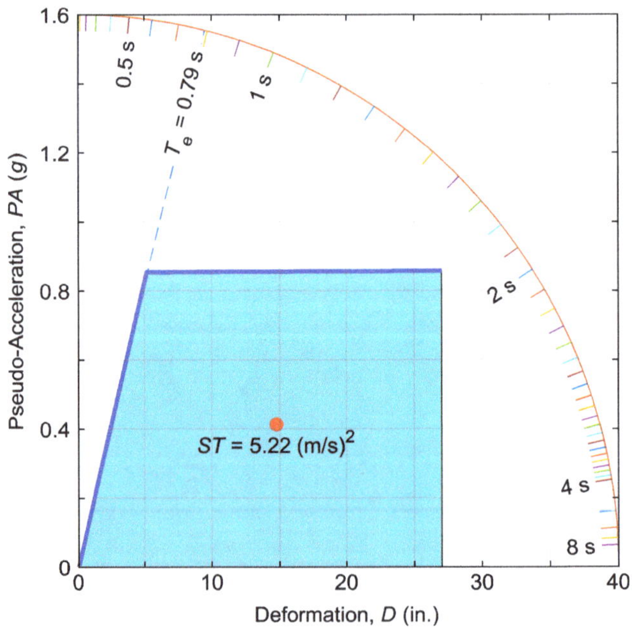

Figure 3-8. Seismic toughness of the 3-story steel special moment frame building as shown in Figure 3-1.

36 moment connections, each with a moment capacity of $M_p = 20{,}790$ kip-in. and rotational capacity of $\theta_p^b = 0.0844$ radian. Ignoring elastic strain energy, strain hardening, and P-Δ, total energy required to fully yield the building is $36 M_p \theta_p^b = 36 \times 20{,}790 \times 0.0844 = 63{,}168$ kip-in. Dividing this by the mass of the building gives $ST \approx \dfrac{63{,}168}{3{,}300/386} \approx 7{,}389\,(\text{in./s})^2 \approx 4.8\,(\text{m/s})^2$, which is 9% smaller than the actual value of 5.22 $(\text{m/s})^2$.

3.5 Damping Curve for 3-Story Building

During elastic response (i.e., before yielding), the building is assumed to possess viscous damping of $\zeta_e = 0.02$ (2% of critical). After yielding, hysteresis becomes the primary source of damping. Figure 3-9 shows the cyclic force–deformation relationship for the building subjected to deformation cycles of various amplitudes. The force–deformation relationship is hysteretic (i.e., energy enclosed within the force–deformation loop is lost during each cycle). As the building continues to yield, hysteretic damping becomes more and more important compared to viscous damping. The overall damping, as a function of deflection $D \geq D_e$, is given by Equation (3-8).

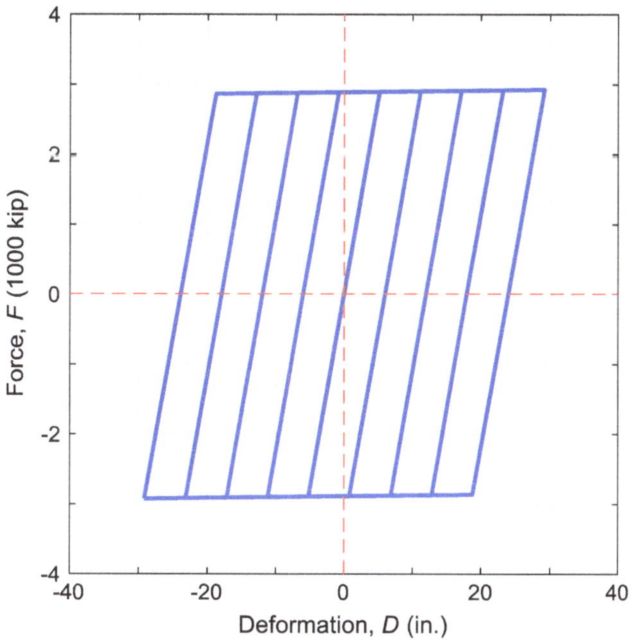

Figure 3-9. Cyclic force–deformation relationship for the building.

$$\zeta = \frac{\zeta_e}{D/D_e}\left(3 - \frac{2}{\sqrt{D/D_e}}\right) + \frac{2\beta}{\pi\alpha(1-\alpha)(D/D_e)}\log_n\left(\frac{1-\alpha+\alpha D/D_e}{(D/D_e)^\alpha}\right) \quad \text{for } \alpha > 0$$

$$\zeta = \frac{\zeta_e}{D/D_e}\left(3 - \frac{2}{\sqrt{D/D_e}}\right) + \frac{2\beta}{\pi}\left(1 - \frac{1+\log_n(D/D_e)}{D/D_e}\right) \quad \text{for } \alpha = 0$$

(3-8)

where α is the post-/pre-yield stiffness, and β is the ratio between the area of a realistic hysteresis loop and that of an ideal hysteresis loop as illustrated in Figure 3-10. The proof of Equation (3-8) is presented in the appendix.

Per Figure 3-6, post-/pre-yield stiffness $\alpha = 0.00087$. It is assumed that $\beta = 0.7$. Figure 3-11 shows a plot of damping versus deformation per Equation (3-8). This is the damping curve for the building. Before yielding, damping is 0.02 (2% of critical). After yielding, the damping increases with deflection to a maximum of 0.23 (23% of critical) at collapse. For this building, 5% damping occurs only at a deflection of 8 in. Damping is <5% (of critical) below 8 in. deflection and >5% (of critical) above 8 in. deflection. That is why 5% assumed damping in prescriptive seismic design is arbitrary.

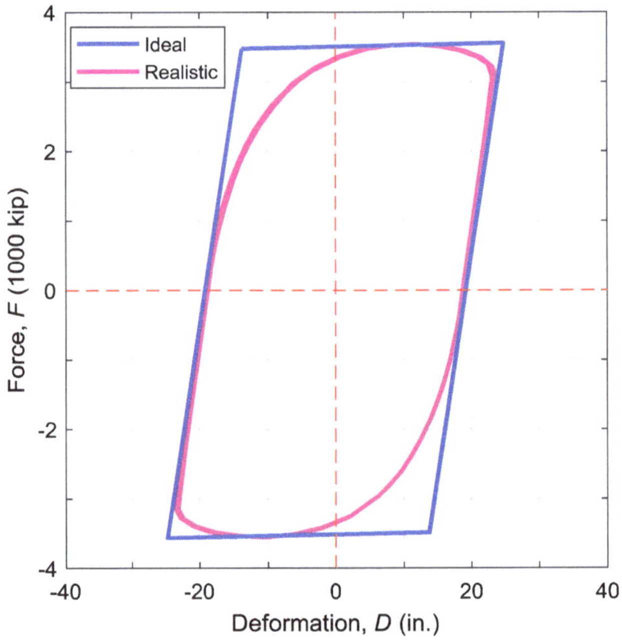

Figure 3-10. Ideal versus realistic hysteresis loop.

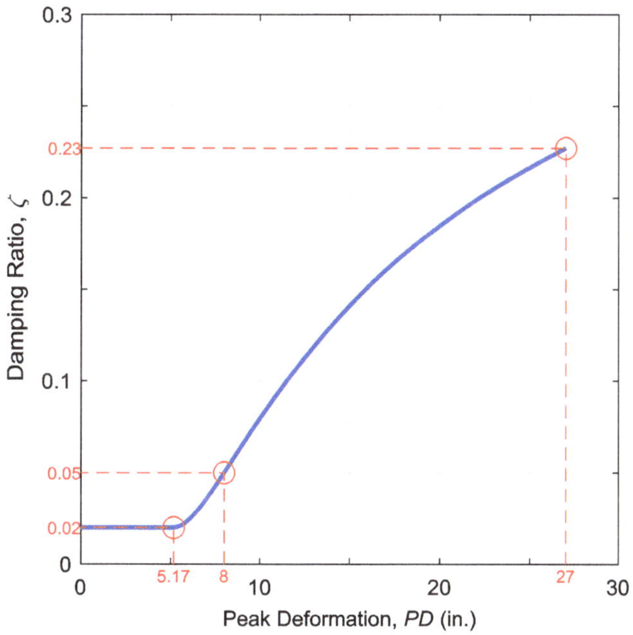

Figure 3-11. Damping curve for the 3-story steel special moment frame building shown in Figure 3-1.

3.6 Twenty-Story Steel Eccentrically Braced Frame Building

Special moment frame is an excellent system for low- to mid-rise buildings because of its high deformability and damping. With increase in height, both elastic and inelastic deflections of the SMF increase. Greater elastic deflection reduces the overall damping of the system because damping in the elastic range is low (2% of critical). Greater total deflection increases the P-Δ effect, and it also requires greater clearance around the building to accommodate ultimate deflection. Therefore, for high-rise buildings, a stiffer system is preferrable. An eccentrically braced frame (EBF) system (Engelhardt and Popov 1989) trades deformability for strength while maintaining or exceeding the seismic toughness, ST (Malhotra 2021). Therefore, an EBF system is selected for the 20-story building.

Figure 3-12 illustrates the EBF system for a 1-story building (Malhotra 2021). The presence of braces reduces the elastic deflection of the building by making it stiffer. For the sake of clarity, elastic deflections are not shown in Figure 3-12. Braces force the plastic rotations to occur away from the columns—at the ends of the link-beam of length e. The shift in plastic rotations away from columns reduces the plastic deflection of the building but increases its lateral strength (Malhotra 2021). The plastic drift ratio θ_p is no longer equal to the plastic rotation in beams θ_p^b, but only a fraction of it:

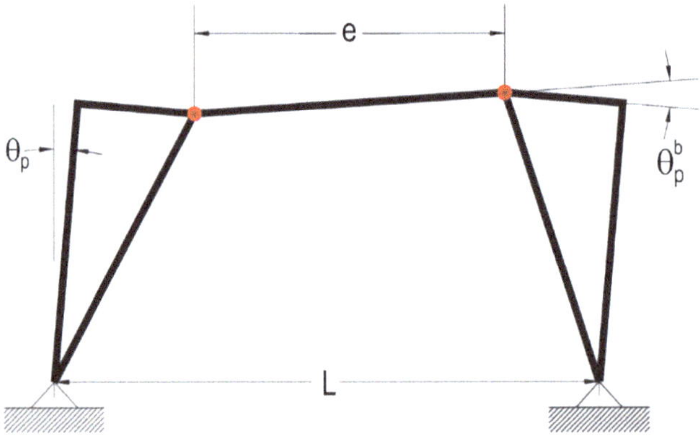

Figure 3-12. Eccentrically braced frame that yields in flexure.
Note: Elastic deformations are not shown in this figure.

$$\theta_p = \theta_p^b \cdot \frac{e}{L} \tag{3-9}$$

With reduction in link-beam length e, the lateral strength of the building increases but its deformability reduces while the seismic toughness ST remains about the same (Malhotra 2021). Reduction in e also increases the shear in the link beam. At a certain optimum value of e, the link-beam yields in shear rather than flexure. This is desirable because the maximum shear rotation in the link-beam is higher than the maximum flexural rotation. Higher value of shear rotation causes an increase in the seismic toughness ST of the building.

Figure 3-13 shows an EBF that yields in shear. The theoretical value of link-beam length for shear yielding is $e \leq 2M_p/V_p$, where M_p and V_p are plastic moment and

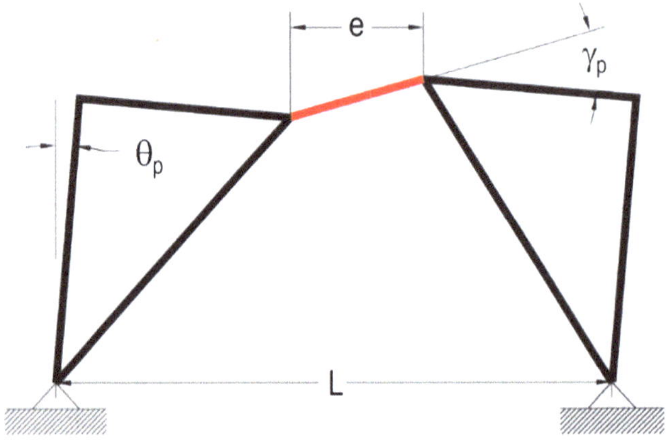

Figure 3-13. Eccentrically braced frame that yields in shear.
Note: Elastic deformations are not shown in this figure.

plastic shear capacities of the link-beam cross section. Design standards such as AISC 341-16 (AISC 2016b) require a more conservative value of link-beam length for shear yielding:

$$e \leq 1.6 M_p / V_p \tag{3-10}$$

The plastic drift ratio of the building θ_p is related to the plastic shear rotation of the link beam by the following expressions:

$$\theta_p = \gamma_p \cdot \frac{e}{L} \tag{3-11}$$

where θ_p, γ_p, e, and L are defined in Figure 3-13.

Figure 3-14 shows a sketch of the 20-story steel building with EBF along the perimeter to resist lateral loads. The footprint and story heights of this building are the same as those of the 3-story building described earlier. There are two EBFs in each orthogonal direction. At each level, EBFs are tied together by concrete-on-steel-deck floor slab which acts as a "rigid" diaphragm. The geometry of the EBF is such that only the link beams (shown in blue color in Figure 3-14) yield in shear while the rest of the EBF remains elastic (AISC 2016b). Link beams are present only on alternate floors. This configuration is also known as a two-story-X

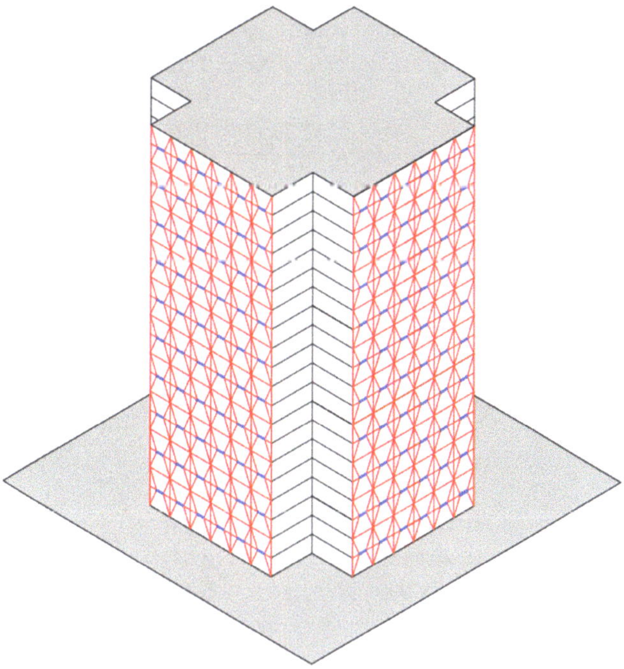

Figure 3-14. Twenty-story steel eccentrically braced frame building.

EBF (Engelhardt and Popov 1989). There are a total of 120 link beams in each orthogonal direction. The total weight of the building and expected live load is $W = 25{,}640$ kip, the total height of the building is $H = 2{,}880$ in., and the height of its CG is $H_{CG} = 1{,}477$ in. Column cross sections are W36 × 439 in 1 to 4 stories, W33 × 318 in 4 to 8 stories, W33 × 221 in 8 to 12 stories, W30 × 191 in 12 to 16 stories, and W30 × 173 in 16 to 20 stories. Column splices occur at midheights of the fourth, eighth, 12th, and 16th stories. Link-beam cross sections vary with height; they are listed in Table 3-1. Structural steel is ASTM A992 with minimum yield stress of $F_y = 50$ ksi and expected yield-stress of $F_{ye} = 55$ ksi. The first-mode period of the building from linear-elastic analysis (CSI 2011) is $T_1 = 1.82$ s. The expected plastic-shear capacity of the link beam (AISC 2016b, ASCE 2017) is

$$V_p = 0.6 F_{ye} \cdot A_s \tag{3-12}$$

where, A_s is the shear area of the link beam in the vertical direction, and A_s and V_p are listed in the third and fourth columns of Table 3-1. To ensure that the link beams yield in shear rather than flexure, their length is kept at $e = 53$ in. per Equation (3-10).

3.7 Pushover Curve for 20-Story Building

As the building is pushed laterally, the link beams start yielding one by one. Figure 3-15 shows plastic shears in the fully yielded EBF. Disregarding the effect of gravity, the lateral force required at the CG to fully yield the building is

$$F_p = \frac{\sum V_p \cdot L}{H_{CG}} \tag{3-13}$$

Table 3-1. Expected Plastic-Shear Capacities of Link-Beams at Various Levels.

Level	Beam cross section	A_s (in.)^2	V_p (kip)
1	W33 × 221	26.296	867.8
3	W33 × 221	26.296	867.8
5	W30 × 191	21.783	718.8
7	W30 × 191	21.783	718.8
9	W30 × 132	18.641	615.2
11	W30 × 132	18.641	615.2
13	W30 × 99	15.418	508.8
15	W30 × 99	15.418	508.8
17	W16 × 89	8.794	290.2
19	W16 × 89	8.794	290.2

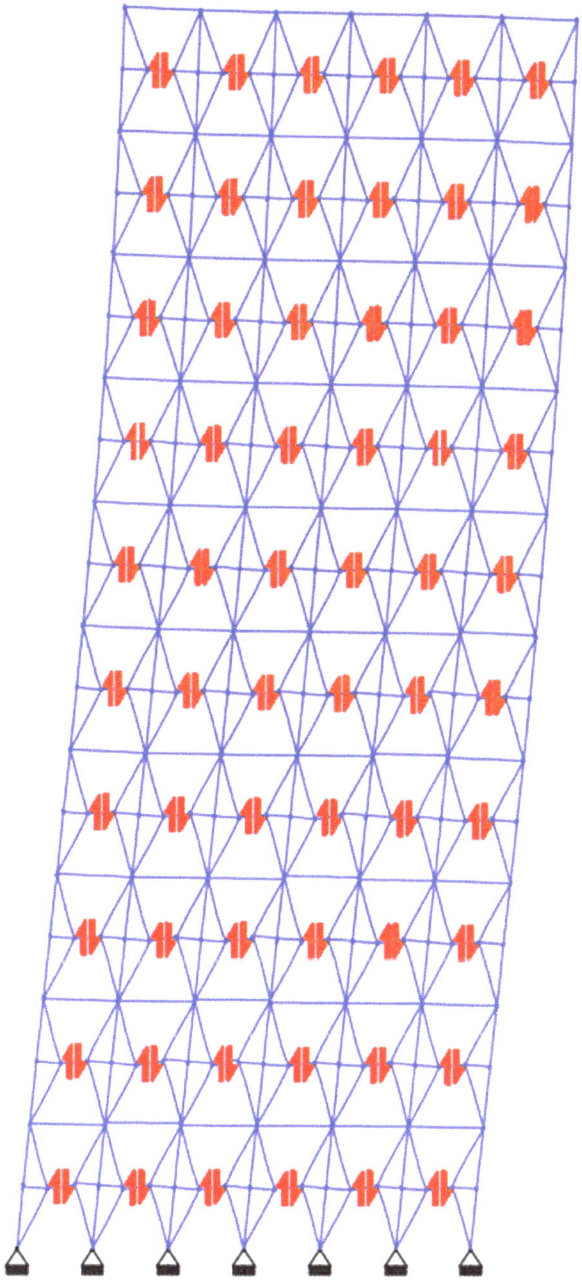

Figure 3-15. Plastic shears in fully yielded eccentrically braced frame building.

where V_p values are per Table 3-1; L = span length (distance between column-centerlines) = 180 in., and H_{CG} = height of the CG of the building = 1,477 in. Upon substituting V_p, L, and H_{CG} into Equation (3-13), F_p = 8,780 kip is obtained.

The elastic stiffness of the building, before the onset of yielding, is given by Equation (3-3). Substituting $W = 25{,}640$ kip and $T_1 = 1.82$ s into Equation (3-3), the elastic stiffness is obtained as $k = 792$ kip/in. The maximum elastic deflection is obtained from Equation (3-4) by substituting $F_p = 8{,}780$ kip and $k = 792$ kip/in.; it is $D_e = 11.1$ in. The elastic portion of the pushover curve ($D \leq D_e$) is generated by using Equation (3-5). For deformation $D \leq D_e$, the force increases linearly with D, but the plastic rotation in link beams γ_p remains zero. For $D > D_e$, the force remains fixed at $F = F_p$, but γ_p increases with increase in deformation.

Per ASCE 41-17, the limiting value of γ_p is 0.16 radian. Substituting $e = 53$ in., $L = 180$ in., and $\gamma_p = 0.16$ radian into Equation (3-11), $\theta_p = 0.0471$ radian is obtained. Substituting $\theta_p = 0.0471$ radian, $D_e = 11.1$ in., and $H_{CG} = 1{,}477$ in. into Equation (3-6), the ultimate deflection of the CG is obtained to be $D_{ult} = 11.1 + 0.0471 \times 1477 = 80.7$ in. The ultimate drift ratio (elastic plus plastic) for the 20-story steel EBF building is $\theta_{ult} = D_{ult}/H_{CG} = 0.0546$ radian. Figure 3-16 shows the simplified pushover curve for the building. The ductility ratio for the 20-story building is $\mu = D_{ult}/D_e = 80.7/11.1 = 7.3$. This is higher than that of the 3-story building.

The pushover curve of Figure 3-16 assumes that all link-beams in the building yield simultaneously. It can be shown through analysis that the building starts yielding at a lateral force of 3,620 kip (and a lateral deflection of 4.6 in.) when the link-beams in the middle bays of ninth level start yielding. With increase in

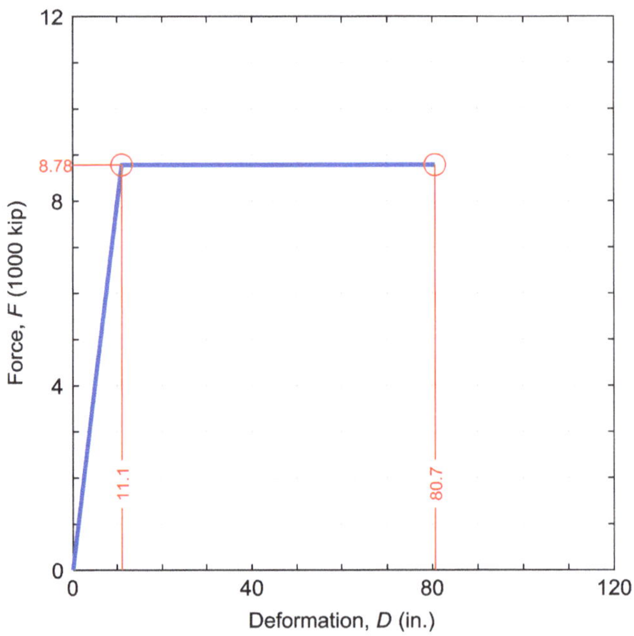

Figure 3-16. Simplified pushover curve for the 20-story steel eccentrically braced frame building.

deflection, the yielding spreads to other link-beams in the building, until the building reaches a fully yielded state. Figure 3-17 compares the ideal pushover curve with a realistic pushover curve. The gradual yielding of the building has some effect on its damping because a realistic hysteresis loop is smaller than an ideal hysteresis loop. Other than that, the use of ideal pushover curve does not introduce significant approximation in the analysis.

The pushover curve of Figure 3-16 ignores any increase in plastic shear because of strain hardening. The pushover curve also ignores the effect of gravity, aka the P-Δ effect (Paulay 1978). The pushover curve is refined as follows.

Strain hardening. Due to strain hardening, the plastic shear V_p increases with increase in shear rotation γ_p. For EBF, AISC 341-16 suggests a 25% increase in shear strength due to strain hardening. Strain-hardening causes the ultimate lateral strength of the building to increase from 8,780 to 10,970 kip. Figure 3-18 shows the pushover curve with strain hardening. The fully yielded, strain-hardened shear capacity of link beams at first level is $867.8 \times 1.25 = 1{,}085$ kip. Connections should remain intact, and columns and braces should be designed to remain elastic for fully yielded, strain-hardened link beams.

P-Δ effect. Referring to Figure 3-15, note that the weight of the structure applies an additional moment that is in the same direction as the moment applied by the

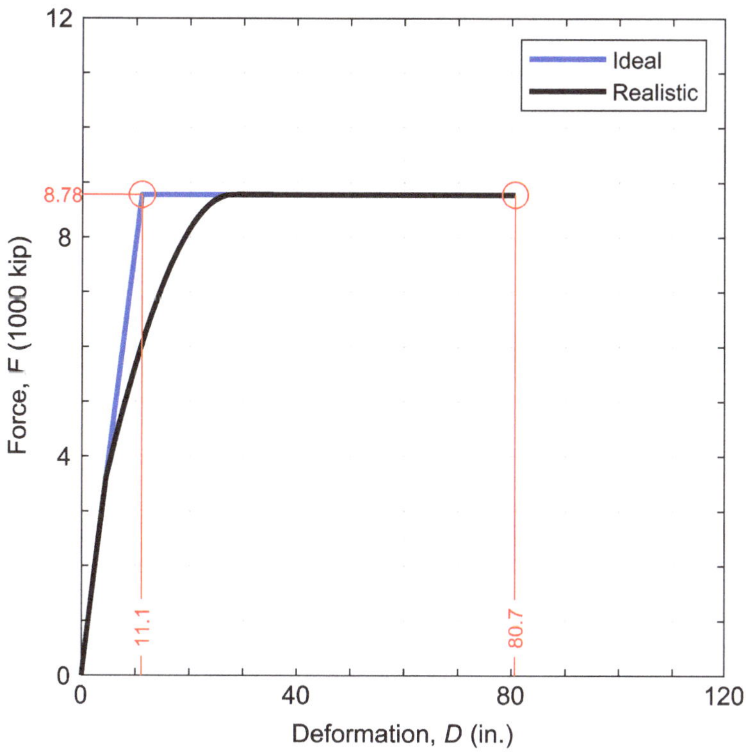

Figure 3-17. Ideal versus realistic pushover curves for the 20-story steel eccentrically braced frame building.

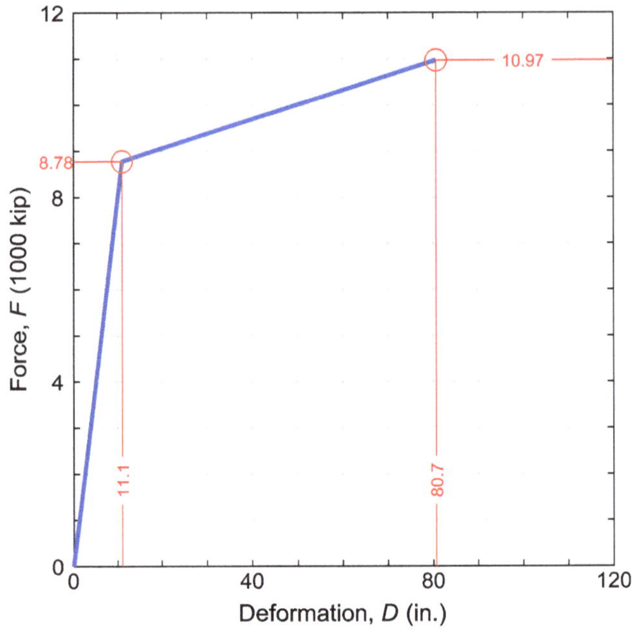

Figure 3-18. Pushover curve for the 20-story steel eccentrically braced frame building with strain hardening.

lateral force F. Both moments are resisted by the EBF. Hence, the weight of the structure reduces the lateral force F for a given deflection D. The P-Δ effect is considered as follows: (1) for a given value of D, F is read from Figure 3-18, and (2) F is reduced by an amount DW/H_{CG}, where W is the weight of the building and H_{CG} is the height of its CG. Figure 3-19 shows the refined pushover curve after considering strain hardening and P-Δ. Because of P-Δ, the yield strength of the building reduces from 8,780 to 8,590 kip, and the ultimate strength of the building reduces from 10,970 to 9,570 kip. The ultimate strength of 9,570 kip is 11% higher than the yield strength of 8,590 kip.

It is important for the pushover curve to have a positive slope (i.e., the lateral force should increase with increase in deflection). If the pushover curve has zero or negative slope, even a slight torsional eccentricity will cause EBF on one side of the building to yield and the EBF on the opposite side of the building to remain elastic. This would cause the building to have a large torsional response, thereby reducing its deformation capacity (Malhotra 2021).

3.8 Capacity Curve for 20-Story Building

The pushover curve of Figure 3-19 is converted to the capacity curve by dividing the force along the vertical axis by the seismic mass $M = 25{,}640$ kip/g. Figure 3-20 shows a

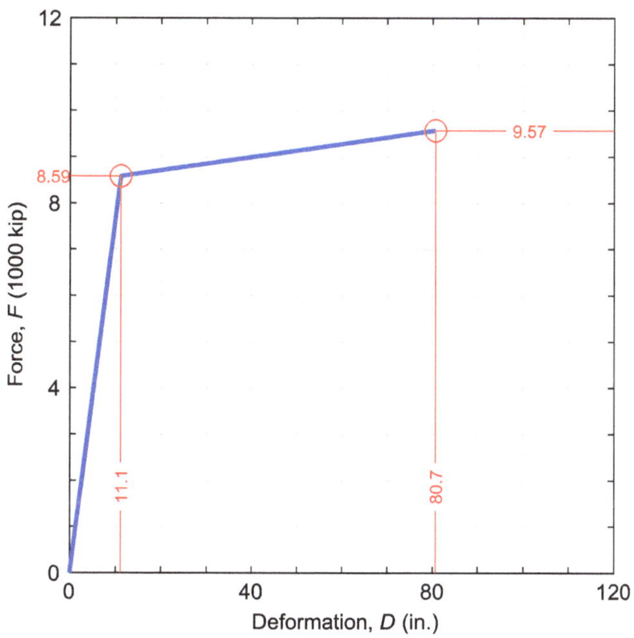

Figure 3-19. Pushover curve for the 20-story steel eccentrically braced frame building with strain hardening and P-Δ.

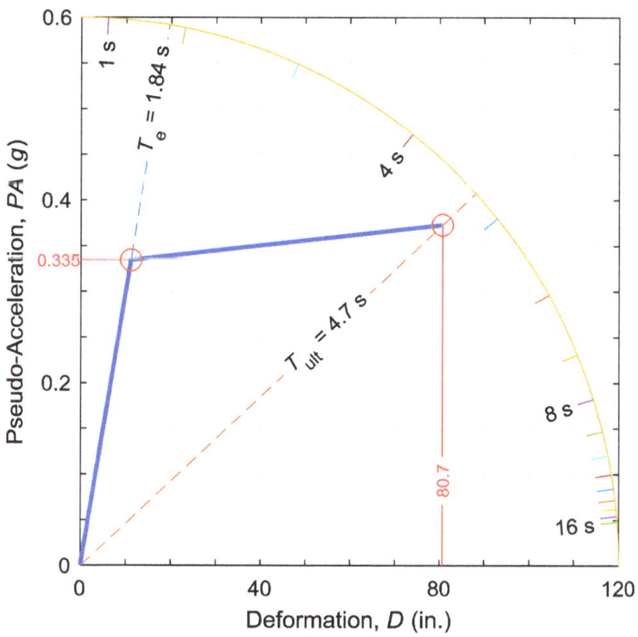

Figure 3-20. Capacity curve for the 20-story steel eccentrically braced frame building shown in Figure 3-14.

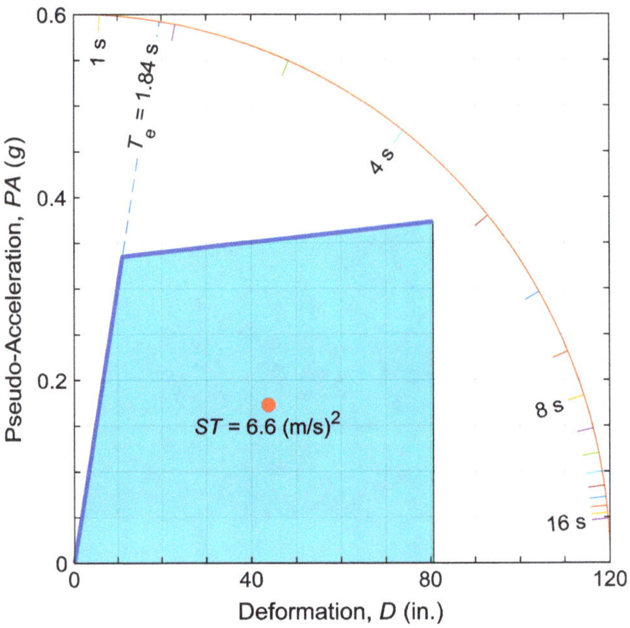

Figure 3-21. Seismic toughness for the 20-story steel eccentrically braced frame building shown in Figure 3-14.

plot of the capacity curve for the 20-story steel EBF building. The vertical axis in Figure 3-20 is the pseudo-acceleration PA in g. The period of the building is 1.84 s at the onset of yielding in link beams at 11.1 in. deflection. After yielding, the period increases with increase in deflection. The capacity curve ends at 80.7 in. deflection when the plastic drift ratio reaches its limiting value of 0.0471 radian and the building collapses. The effective period of the building at collapse is $T_{ult} = 4.7$ s. The area below the capacity curve (Figure 3-21) is the seismic toughness ST. For the 20-story steel EBF building, $ST = 6.6\,(\text{m/s})^2$, which is 26% higher than that of the 3-story steel SMF building.

As a sanity check, approximate value of ST is obtained by adding the toughness of all structural elements. There are 120 link beams, each with a rotation capacity of $\gamma_p = 0.16$ radian and shear capacities V_p listed in the last column of Table 3-1. The work done to yield each link beam is $V_p \times e \times \gamma_p$. The total work done to yield all 120 link beams is $\sum V_p \times e \times \gamma_p = 610{,}720$ kip-in. Dividing this by the mass of the building gives $ST \approx \dfrac{610{,}720}{25{,}640/386} \approx 9{,}194\,(\text{in./s})^2 \approx 5.9\,(\text{m/s})^2$, which is 10% smaller than the actual value of $6.6\,(\text{m/s})^2$.

3.9 Damping Curve for 20-Story Building

The damping curve for the building is generated per Equation (3-8). According to Figure 3-19, post-/pre-yield stiffness $\alpha = 0.0183$. It is again assumed that $\beta = 0.7$.

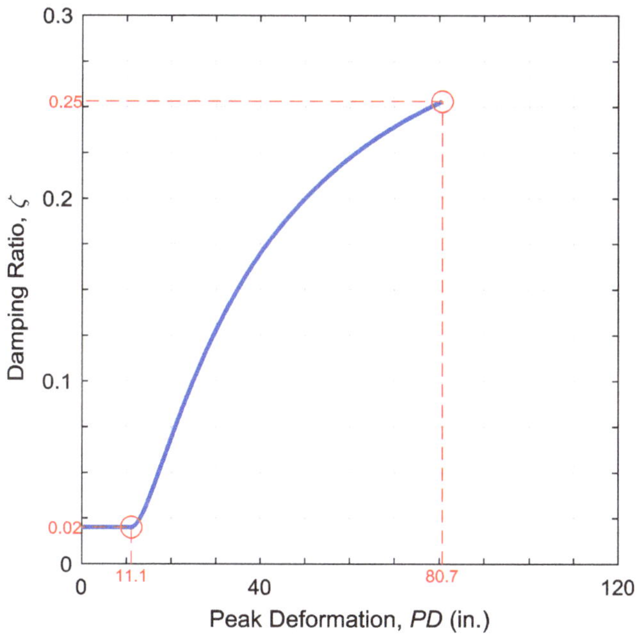

Figure 3-22. Damping curve for the 20-story steel eccentrically braced frame building shown in Figure 3-14.

Figure 3-22 shows a plot of the damping versus deformation for the 20-story steel EBF building per Equation (3-8). For this building, 5% damping occurs at a deflection of 17.2 in. Damping is <5% (of critical) below 17.2 in. deflection and >5% (of critical) above 17.2 in. deflection.

3.10 Comparison between 3- and 20-Story Buildings

Figure 3-23 compares the capacity curves for 3- and 20-story buildings. The 3-story building has higher normalized strength, but it is less deformable than the 20-story building. The area below the capacity curve is slightly higher for the 20-story building, implying that the 20-story building has higher seismic toughness ST than the 3-story building. The ultimate period of the 3-story building is $T_{ult} = 1.79$ s, and that of the 20-story building is $T_{ult} = 4.7$ s. To collapse the 3-story building, ground motions will have to be rich in intermediate frequencies ($0.5\,\text{s} < T < 2\,\text{s}$). To collapse the 20-story building, ground motions will have to be rich in low frequencies ($T > 2\,\text{s}$). Note that only stiff and brittle structures are collapsed by high frequencies.

The ductility ratio μ is the ratio between the ultimate deflection and the elastic deflection. Structures with higher ductility ratio μ undergo more plastic deformation compared to elastic deformation. Damping associated with elastic deformation is low

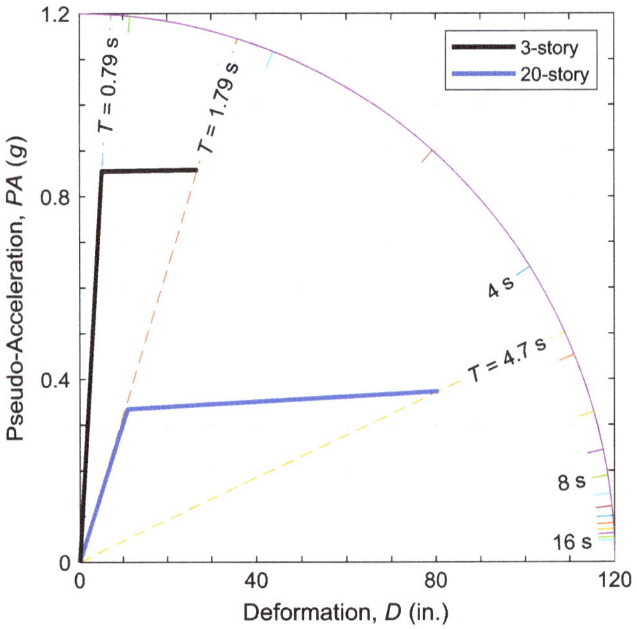

Figure 3-23. Capacity curves for 3-story and 20-story buildings.

(2% of critical). Therefore, structures with higher μ have greater damping. The ductility ratio for the 3-story building is $\mu = 27.2/5.35 = 5.08$. The ductility ratio for the 20-story building is $\mu = 80.7/11.1 = 7.3$. Higher ductility ratio for the 20-story building is the cause of higher damping for that building. Figure 3-24 compares the damping curves for the two buildings. Because of its higher elastic deformation, the 20-story building must deflect more than the 3-story building to achieve the same damping ratio.

3.11 Cyclic Capacity

The limiting values of plastic rotations θ_p^b and γ_p should also depend on the number of cycles. The greater the number of cycles, the smaller should be the limiting values of θ_p^b and γ_p. However, the number of cycles is not clearly mentioned in seismic guidelines (AISC 2016a, b; ASCE 2017). The commonly used cyclic test protocols per Section K2.4 of AISC 341-16 (AISC 2016b) are shown in Figures 3-25 and 3-26. Based on these test protocols, it is conservatively assumed that the limiting value of θ_p^b corresponds to two cycles and the limiting value of γ_p corresponds to one cycle. The CG of the 3-story building can deflect a maximum of 27 in. for two cycles, and the CG of 20-story building can deflect a maximum of 80.7 in. for one cycle.

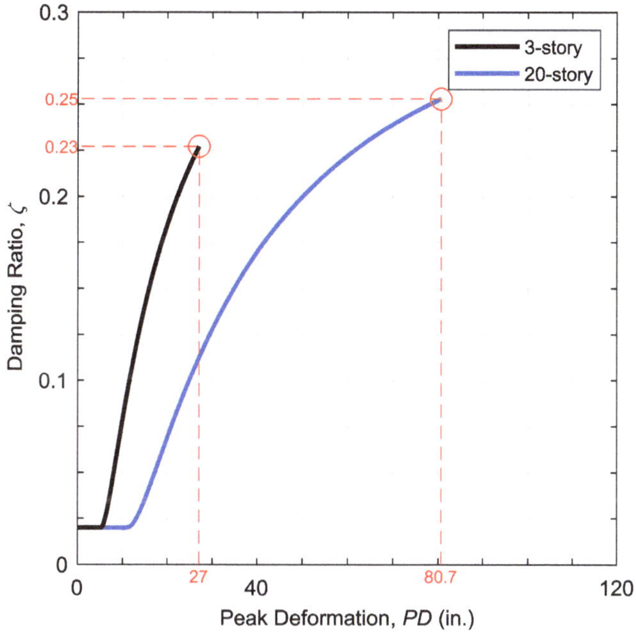

Figure 3-24. Damping curves for 3-story and 20-story buildings.

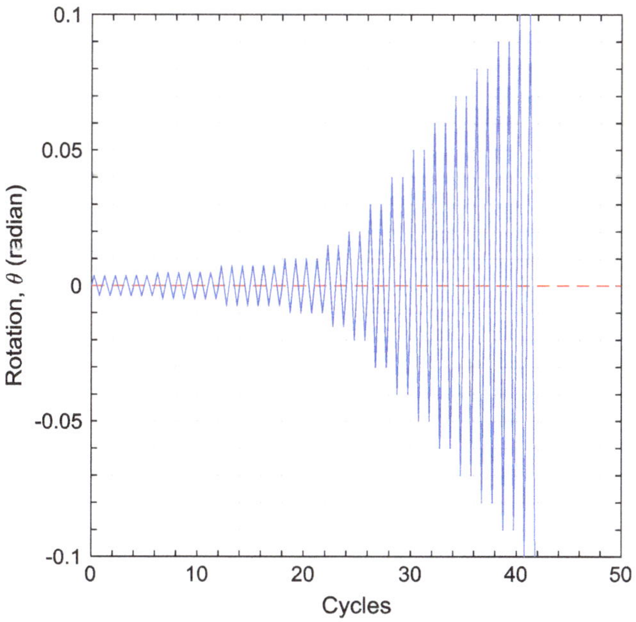

Figure 3-25. Cyclic test protocol for total rotation angle θ in special moment frames per AISC 341-16.

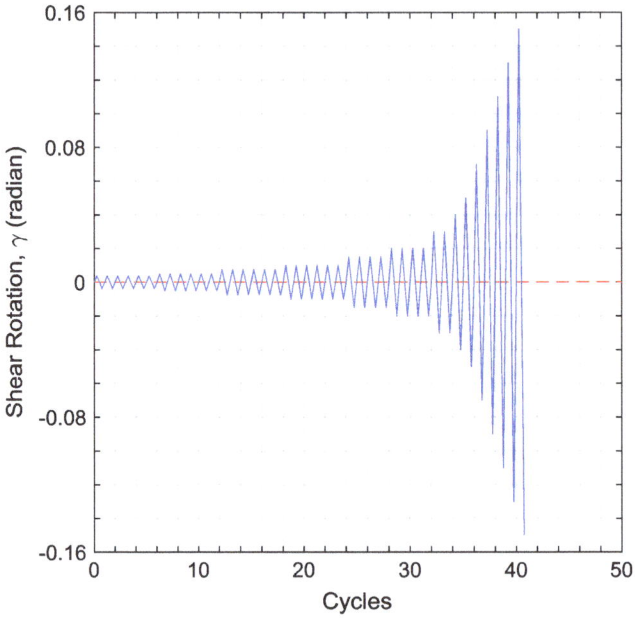

Figure 3-26. Cyclic test protocol for total shear rotation angle γ in eccentrically braced frames as per AISC 341-16.

3.12 Summary

1. Normalized strength, deformability, and damping are the important seismic characteristics of a structure.
2. The capacity curve defines the normalized strength and deformability of a structure. The damping curve defines its energy-dissipation capacity.
3. The area below the capacity curve is seismic toughness ST of the structure. ST captures the effects of normalized strength and deformability; it can be approximately determined by adding the toughness of structural elements that are expected to yield.
4. The period of a structure is not fixed; it changes during the ground shaking. The period of a yielded structure can be significantly longer than that of an elastic structure.
5. The damping of a structure is also not fixed; it changes during the ground shaking. The damping of a yielded structure can be significantly higher than that of an elastic structure.
6. Hysteresis is the main source of damping after yielding. Structures with fuller hysteresis loops exhibit more damping than those with pinched hysteresis loops.
7. The ductility ratio (defined as the ratio between ultimate and elastic deflection) determines the maximum damping of a structure.

8. A structure can be thought of as a chain with several links. To achieve high deformability, brittle links should be made stronger than ductile links.
9. A structure cannot be considered deformable if it lacks sufficient clearance all around to deform freely.
10. Strain hardening gives a positive slope to the force–deformation relationship after yielding. This is helpful in spreading yielding within the structure to achieve high deformability. Strain hardening needs to be more clearly defined in design guidelines.
11. P-Δ reduces the lateral strength of a structure and it reduces the effect of strain hardening. The deflection of a structure should not be so high that P-Δ causes the slope of the pushover curve to become negative.
12. Cyclic capacity is not sufficiently clarified in current seismic guidelines. There is a need to determine and define plastic rotation limits for different number of cycles: 1, 2, 3, 4, and so on.

References

AISC (American Institute of Steel Construction). 2016a. *Prequalified connections for special and intermediate steel moment frames for seismic applications, including Supplement No. 1.* AISC 358-16, AISC 358s1-18. Chicago: AISC.

AISC. 2016b. *Seismic provisions for structural steel buildings.* AISC 341-16. Chicago: AISC.

ASCE. 2017. *Seismic evaluation and retrofit of existing buildings.* ASCE 41-17. Reston, VA: ASCE.

ASCE. 2022. *Minimum design loads and associated criteria for buildings and other structures.* ASCE 7-22. Reston, VA: ASCE.

CSI (Computer and Structures, Inc.). 2011. *SAP 2000.* Berkeley, CA: CSI.

Engelhardt, M. D., and E. P. Popov. 1989. "Behavior of long links in eccentrically braced frames." *UCB/EERC-89/01.* Berkeley, CA: Earthquake Engineering Research Center.

Hamburger, R. O., H. Krawinkler, J. O. Malley, and S. M. Adan. 2009. "Seismic design of steel special moment frames: A guide for practicing engineers." NEHRP Seismic Design Tech. Brief No. 2, *NIST GCR 09-917-3,* Gaithersburg, MD: National Institute of Standards and Technology, US Department of Commerce.

Malhotra, P. K. 2021. *Seismic analysis of structures and equipment.* Cham, Switzerland: Springer.

Paulay, T. 1978. "A consideration of *P*-delta effects in ductile reinforced concrete frames." *Bull. New Zealand Nat. Soc. Earth. Eng.* 111 (3): 151–160.

SEAOC (Structural Engineers Association of California). 2012. *2012 IBC SEAOC structural/seismic design manual.* Vol. 4: Examples for steel-framed buildings. Sacramento, CA: SEAOC.

CHAPTER 4

Seismic Analysis of Buildings

4.1 Introduction

During extreme ground motions listed in Table 1-2, the buildings (described in Chapter 3) are not expected to remain linear elastic. They are expected to yield and respond in a nonlinear fashion. There are two methods of nonlinear analysis (Malhotra 2021): (1) nonlinear-static, and (2) nonlinear-dynamic. Nonlinear-static analysis is intuitive, and it does not require numerical solution of coupled-nonlinear equations of motion (EOM). Hence, it is free of numerical errors. Nonlinear-dynamic analysis requires numerical solution of coupled-nonlinear EOM. Nonlinear-dynamic analysis can be more "accurate" than nonlinear-static analysis, but it can also be easily corrupted by numerical errors. Nonlinear-dynamic analysis is less intuitive because it relies on proprietary software whose inner working are not known to an average engineer.

In this book, nonlinear-static analysis is the primary method of analysis. Nonlinear-dynamic analysis is used only to confirm that the responses are not underestimated. In this chapter, both methods of analysis are used to calculate the responses of 3-story steel SMF and 20-story steel EBF buildings (described in Chapter 3) to the TAPS ground motion discussed in Chapter 1.

4.2 Demand Curve for 3-Story Building

As discussed in Chapter 1, ground motions impose both amplitude and cyclic demands on structures. The former is expressed by the ADRS, and the latter is expressed by the cyclic-demand response spectra (CDRS). Figures 4-1 and 4-2, respectively, show the ADRS and CDRS of the TAPS ground motion for various values of damping.

Figure 4-3 shows the damping curve for the building (previously shown in Figure 3-11). The viscous damping of the building in elastic range is $\zeta_e = 0.02$ (2% of critical). With increase in deformation, the damping increases because of hysteresis. The damping is 23% of critical at ultimate deflection of $D_{ult} = 27$ in. The TAPS ground motion demand-curve adjusted for the damping of 3-story steel SMF building is generated as follows:

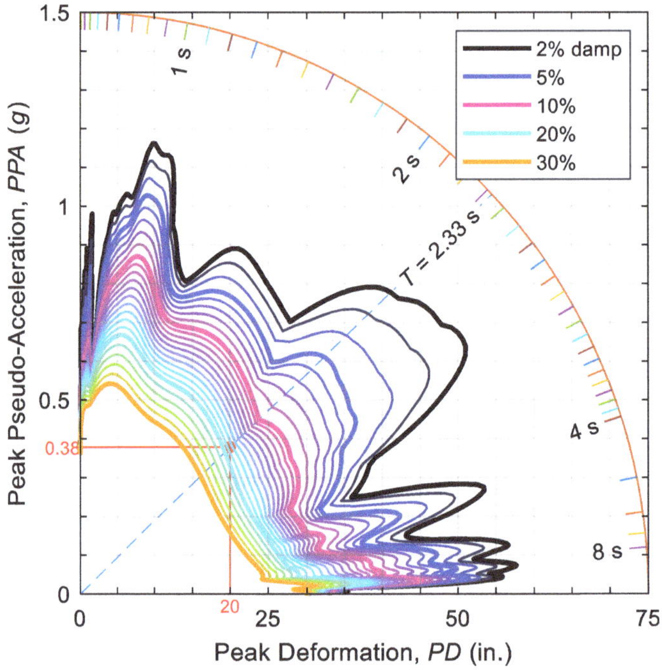

Figure 4-1. Acceleration–deformation response spectra of TAPS ground motion for various values of damping.

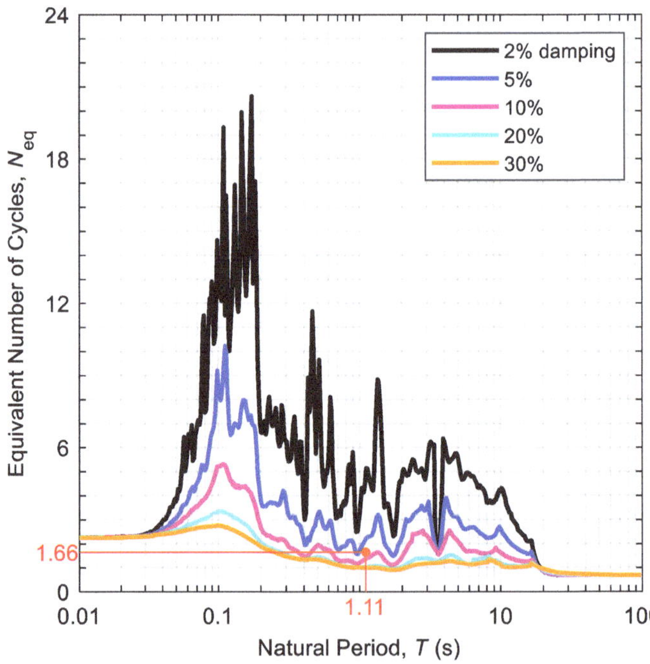

Figure 4-2. Cyclic-demand response spectra of TAPS ground motion for various values of damping.

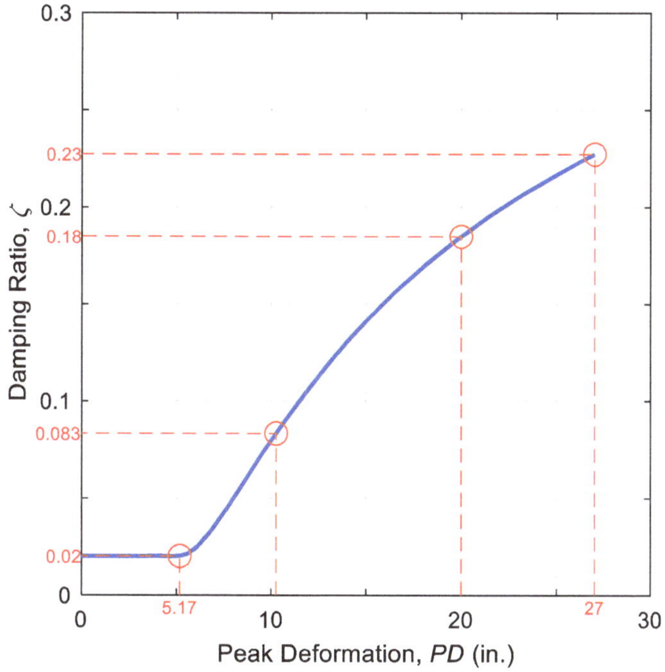

Figure 4-3. Damping curve for the 3-story steel special moment frame building as shown in Figure 3-1.

1. For a given value of peak deformation PD, damping is read from Figure 4-3. For example, for $PD = 20$ in., $\zeta = 0.18$ (18% of critical). For deformations greater than D_{ult}, the damping is assumed same as that for $D = D_{ult}$. Note that the demand curve for $D > D_{ult}$ is not important because the building has already collapsed.
2. Corresponding to PD and ζ in Step 1, the peak pseudo-acceleration PPA is read from the ADRS shown in Figure 4-1. For $PD = 20$ in. and $\zeta = 0.18$, $PPA = 0.38g$ per Figure 4-1. It is possible that for certain values of PD, there are more than one values of PPA. Period is calculated for each PD and PPA pair from the equation: $T = 2\pi\sqrt{PD/PPA}$. Note that T can also be read from the radial tick marks as shown in Figure 4-1. For $PD = 20$ in. and $PPA = 0.38g$, $T = 2.33$ s.
3. PD and PPA pairs are arranged in ascending order of period T. A plot of PPA versus PD is the damping-adjusted demand curve. Figure 4-4 shows the TAPS ground motion demand curve adjusted for damping of the 3-story steel SMF building. Different segments of the demand curve correspond to different values of damping as indicated in Figure 4-4. In comparison with Figure 1-17, notice that the energy demand ED has reduced from 6.98 to 4.15 $(\text{m/s})^2$, and the energy centroid T_{cg} has shifted from 2.2 to 1.68 s. The reduction in ED is the reduced demand owing to damping of the building. The reduction in T_{cg} is because of a greater reduction in demand at higher deflections (or longer periods).

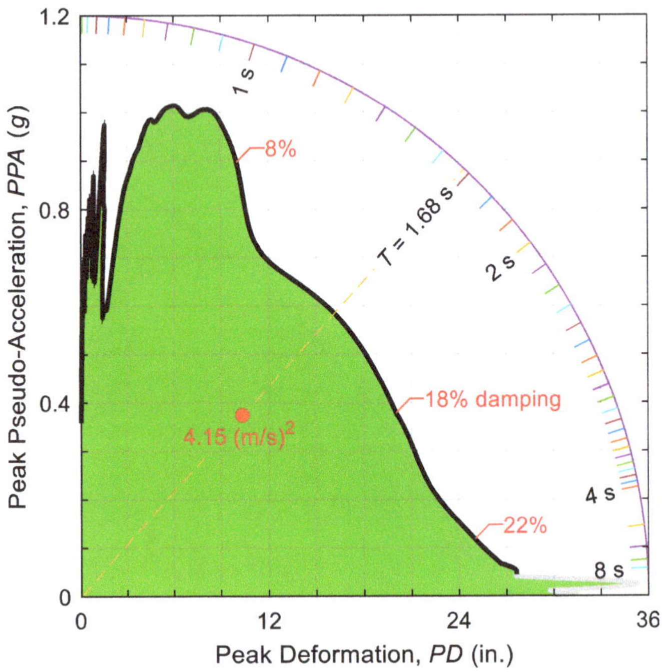

Figure 4-4. TAPS ground motion demand curve adjusted for damping of 3-story steel special moment frame building.

4.3 Nonlinear-Static Analysis of 3-Story Building

Per Figure 3-8, the seismic toughness of the building is $ST = 5.22\ (\text{m/s})^2$. Because $ED = 4.15\ (\text{m/s})^2$ is less than ST, the building cannot collapse, but it can be damaged by the TAPS ground motion. To determine the amount of damage, the response of the building is determined from the intersection of demand and capacity curves. The capacity curve for the building was generated in Chapter 3. In Figure 4-5, the capacity curve of Figure 3-7 is superimposed on the demand curve of Figure 4-4. The demand-capacity equilibrium occurs at a deflection of $D_{eq} = 10.3$ in. The effective period of the building at equilibrium is $T_{eq} = 1.11$ s per Figure 4-5. The effective damping of the building corresponding to $D_{eq} = 10.3$ in. is $\zeta_{eq} = 0.083$ per Figure 4-3. The equivalent number of cycles corresponding to $T_{eq} = 1.11$ s and $\zeta_{eq} = 0.083$ is $N_{eq} = 1.66$ per Figure 4-2. Therefore, the CG of the building deflects 10.3 in. for 1.66 cycles during the TAPS ground motion. The total (elastic + plastic) drift ratio of the building during the TAPS ground motion is $10.3/H_{CG} = 10.3/259 = 0.0396$ radian. The deflection of the roof is $0.0396 \times H = 0.0396 \times 432 = 17.1$ in. The first row of Table 4-1 lists the responses of the building from the nonlinear-static analysis.

Figure 4-6 shows a plot of "energies" that are absorbed and repelled by the building during the TAPS ground motion. The energy demand for 5% assumed

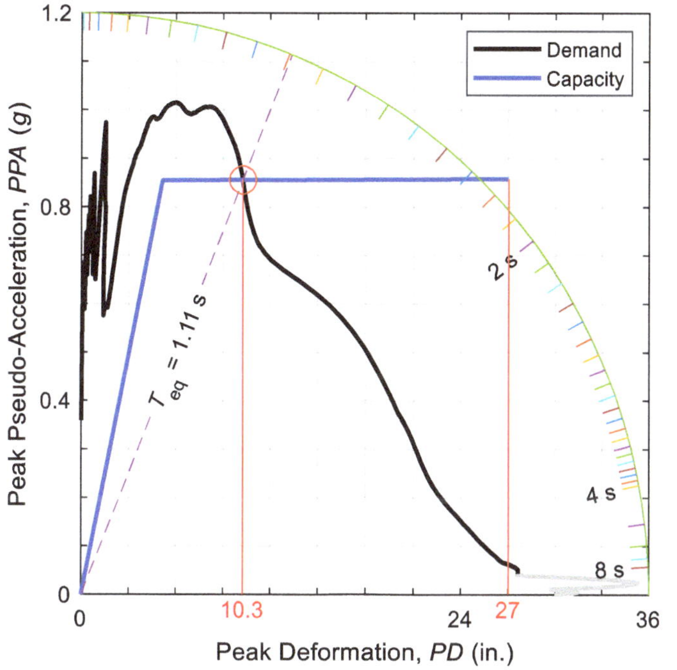

Figure 4-5. Demand-capacity equilibrium of 3-story steel special moment frame building during the TAPS ground motion.

Table 4-1. Responses of 3-Story Steel Special Moment Frame Building During the TAPS Ground Motion.

Nonlinear analysis	Roof deflection (in.)	Maximum drift ratio (radian)	Number of cycles	Fatigue damage (%)
Static	17.1	0.0396	1.66	12
Dynamic	10.8	0.0335	0.76	4

damping is 6.98 $(m/s)^2$ (Figure 1-17). Owing to damping of the building, *ED* reduces to 4.15 $(m/s)^2$ (Figure 4-4), out of which 1.64 $(m/s)^2$ is absorbed, and 2.51 $(m/s)^2$ is repelled by the building. Most of the repelled energy is the low-frequency energy to the right of the absorbed energy in Figure 4-6.

4.4 Nonlinear-Dynamic Analysis of 3-Story Building

A nonlinear model of the building is generated in SAP 2000 (CSI 2011). Beams are modeled by Wen-Bouc (Wen 1976) plasticity element with exponent $\eta = 0.5$ and post-/pre-yield stiffness ratio $\alpha = 0.00087$ (Figure 3-6). The parameter η in Wen-Bouc (Wen 1976) model defines the shape of the hysteresis loop. The chosen values of η

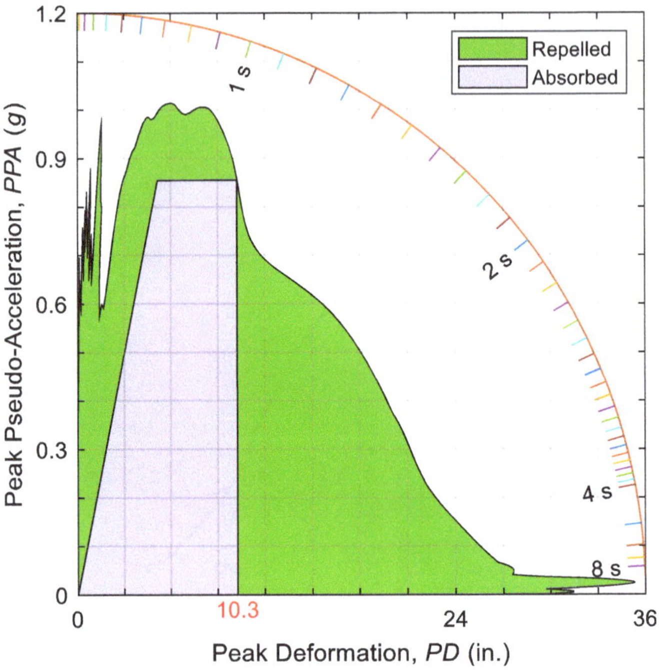

Figure 4-6. Energies absorbed and repelled by the 3-story building during the TAPS ground motion.

and α attempt to match the stiffness and damping characteristics of the building in static and dynamic analyses. The building is subjected to the east and north direction ground motions shown in Figure 1-1. Because of the flexibility of columns, the inter-story drift ratio is not the same in every story. The maximum inter-story drift ratio is observed in the first story. Figure 4-7 shows histories of drift ratios in the east and

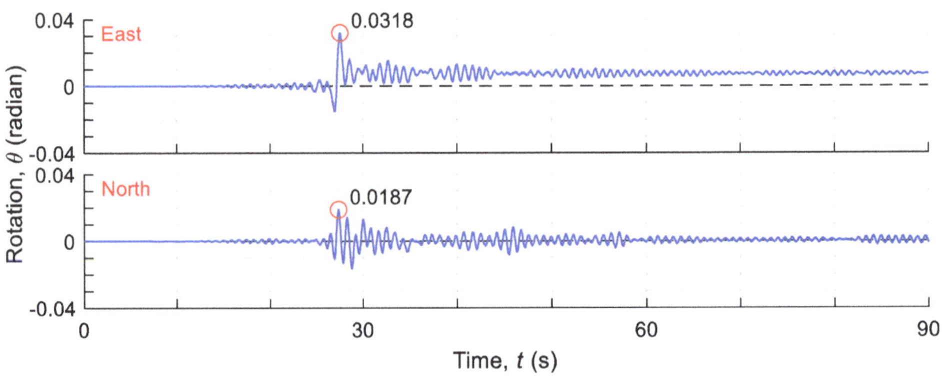

Figure 4-7. First-story drift ratios in 3-story steel special moment frame building during the TAPS ground motion.

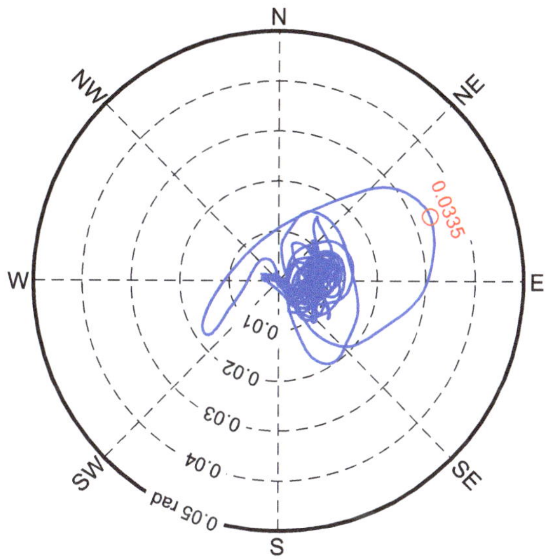

Figure 4-8. Resultant first-story drift ratios in 3-story building during the TAPS ground motion.

north directions. It is likely that the drift ratio can increase by changing the orientation of the building with respect to the ground motion. Therefore, the drift ratios in two orthogonal directions are combined to generate the resultant drift ratios shown in Figure 4-8. The maximum resultant drift ratio is 0.0335 radian, and it occurs in 72 degrees clockwise from north (N72E) direction. Figure 4-9 shows the history of drift ratios in N72E direction.

The equivalent number of cycles is determined as described in Section 1.11. Each "zero-crossing" half-cycle is assigned a weight that is proportional to the square of its amplitude. Elastic perturbations that do not cross the "zero line" are not counted because they do not cause damage. Adding the weights and dividing by 2 gives the equivalent number of cycles to be $N_{eq} = 0.76$. Therefore, per dynamic analysis, peak drift ratio of 0.0335 radian occurs for 0.76 cycles. Figure 4-10 shows resultant roof deflections from dynamic analysis. The maximum roof deflection is 10.8 in. The second row of Table 4-1 lists the responses of the building from nonlinear-dynamic analysis.

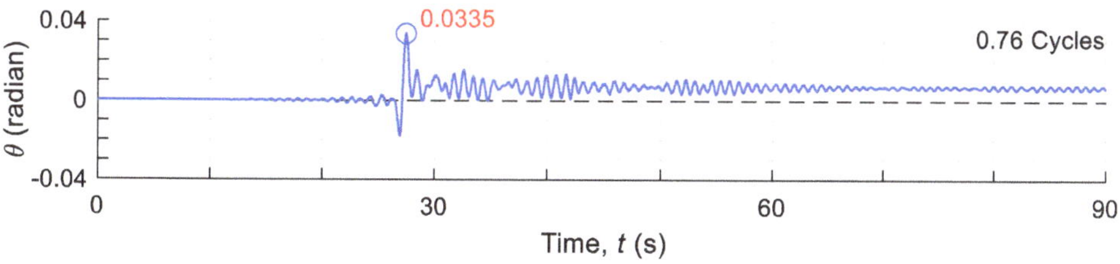

Figure 4-9. First-story drift ratio in N72E direction during the TAPS ground motion.

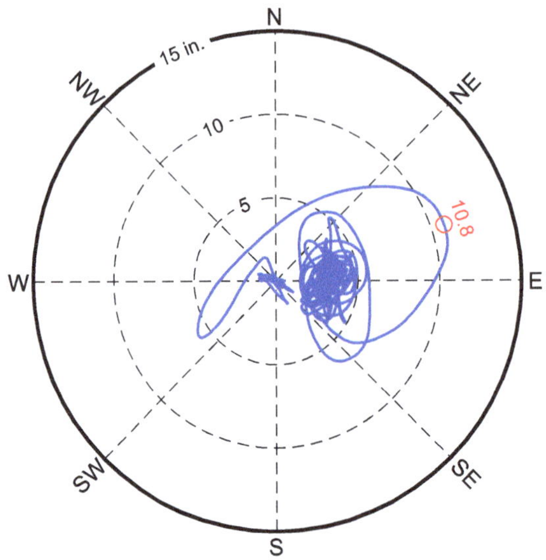

Figure 4-10. Resultant roof deflection of 3-story building during the TAPS ground motion.

4.5 Damage Suffered by 3-Story Building

Per static analysis, the drift ratio is 0.0396 radian for 1.66 cycles, but the maximum drift capacity is $\theta_{ult} = 0.104$ radian for two cycles. Assuming that the damage is proportional to (1) the square of amplitude, and (2) the number of cycles, the structural damage suffered by the building during the TAPS ground motion is $0.0396^2 \times 1.66/(0.104^2 \times 2) = 0.12$ (12%).

Per dynamic analysis, the story drift ratio is 0.0335 radian for 0.76 cycles, but the maximum drift capacity is $\theta_{ult} = 0.104$ radian for two cycles. The structural damage suffered by the building during the TAPS ground motion is $0.0335^2 \times 0.76/(0.104^2 \times 2) = 0.04$ (4%).

From both static and dynamic analyses, the damage suffered by the building during the TAPS ground motion is negligible. Therefore, the building should be able to withstand additional damage from aftershocks without the danger of collapse. Next, the response of 20-story steel EBF building during the TAPS ground motion is examined.

4.6 Demand Curve for 20-Story Building

Although the ground motion is the same, the demand curve for the 20-story steel EBF building is different from that for the 3-story steel SMF building because of the difference in damping. Figure 4-11 shows the damping curve for the 20-story steel EBF building previously generated in Chapter 3. The demand curve for the 20-story

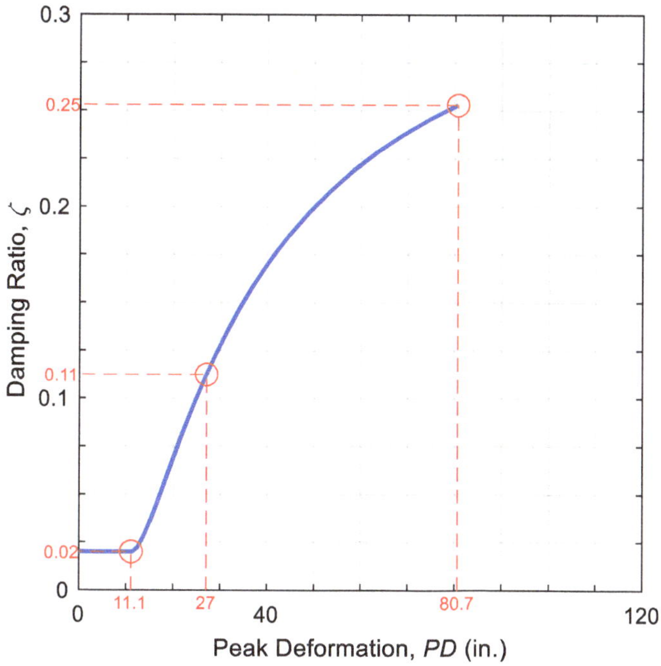

Figure 4-11. Damping curve for 20-story steel eccentrically braced frame building as shown in Figure 3-14.

steel EBF building is generated by using the same steps as described in Section 4.2. It is shown in Figure 4-12. The damping-adjusted energy demand for the TAPS ground motion is $ED = 5.71\ (\text{m/s})^2$. In comparison with Figure 4-4, notice that damping is less effective in reducing the ED for 20-story building. This is because, for the same deflection, damping for the 20-story building is smaller than that for the 3-story building, as seen in Figure 3-24.

4.7 Nonlinear-Static Analysis of 20-Story Building

Because the damping-adjusted $ED = 5.71\ (\text{m/s})^2$ is less than the seismic toughness of $ST = 6.6\ (\text{m/s})^2$ (Figure 3-21), the building cannot collapse, but it can be damaged by the TAPS ground motion. The response of 20-story steel EBF building to TAPS ground motion is determined from the intersection of demand and capacity curves. The capacity curve for the building was generated in Chapter 3. In Figure 4-13, the capacity curve of Figure 3-20 is superimposed on the demand curve of Figure 4-12. The demand-capacity equilibrium occurs at a deflection of $D_{eq} = 27\,\text{in.}$ The effective period of the building at equilibrium is $T_{eq} = 2.84\,\text{s}$ per Figure 4-13. The effective damping of the building corresponding to $D_{eq} = 27\,\text{in.}$ is $\zeta_{eq} = 0.11$ per Figure 4-11. The equivalent number of cycles corresponding to $T_{eq} = 2.84\,\text{s}$ and $\zeta_{eq} = 0.11$ is

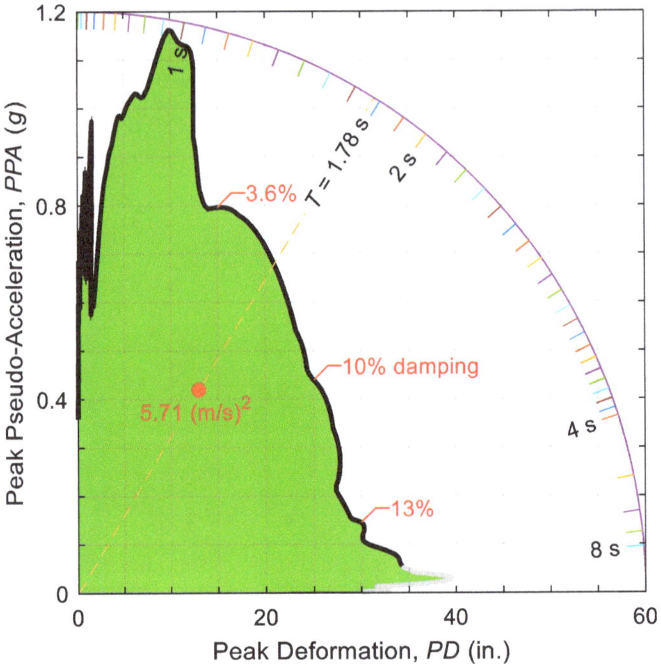

Figure 4-12. TAPS ground motion demand curve adjusted for damping of 20-story steel eccentrically braced frame building.

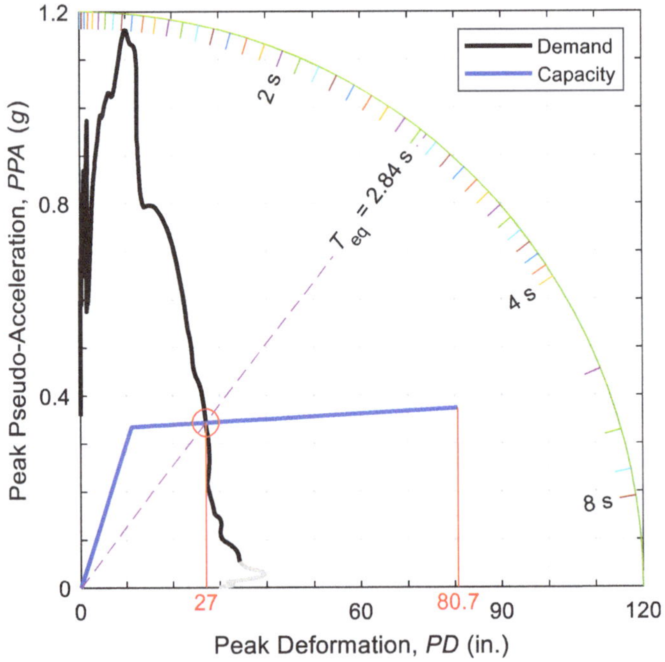

Figure 4-13. Demand-capacity equilibrium of 20-story steel eccentrically braced frame building during the TAPS ground motion.

Table 4-2. Responses of 20-Story Steel Eccentrically Braced Frame Building During the TAPS Ground Motion.

Nonlinear analysis	Roof deflection (in.)	Maximum drift ratio (radian)	Number of cycles	Fatigue damage (%)
Static	52.7	0.0183	2.33	26
Dynamic	30.7	0.0163	1	9

$N_{eq} = 2.33$ per Figure 4-2. Therefore, the CG of 20-story steel EBF building deflects 27 in. for 2.33 cycles during the TAPS ground motion. The total (elastic + plastic) drift ratio of the building during the TAPS ground motion is $27/H_{CG} = 27/1477 = 0.0183$ radian. The deflection of roof is $0.0183 \times H = 0.0183 \times 2880 = 52.7$ in. The first row of Table 4-2 lists the responses of the building from the nonlinear-static analysis.

Figure 4-14 shows a plot of "energies" that are absorbed and repelled by the building during the TAPS ground motion. The energy demand for 5% assumed damping is 6.98 $(m/s)^2$ (Figure 1-17). Owing to the damping of the building, *ED* reduces to 5.71 $(m/s)^2$ (Figure 4-12), out of which 1.81 $(m/s)^2$ is absorbed by the

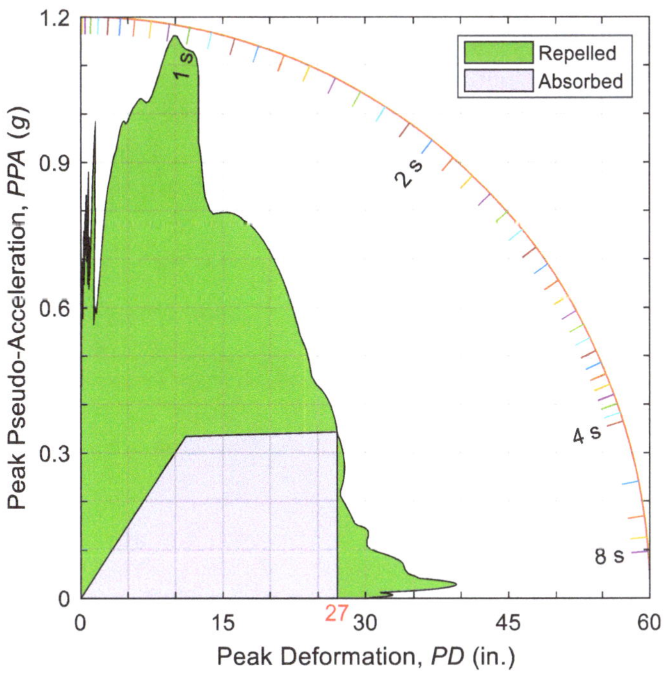

Figure 4-14. Energies absorbed and repelled by the 20-story building during TAPS ground motion.

building and 3.9 $(m/s)^2$ is repelled by the building. Most of the repelled energy is high-frequency energy shown above the absorbed energy in Figure 4-14.

4.8 Nonlinear-Dynamic Analysis of 20-Story Building

A nonlinear model of the building is generated in SAP 2000 (CSI 2011). Link beams are modeled by Wen-Bouc (Wen 1976) plasticity element with exponent $\eta = 0.5$ and post-/pre-yield stiffness ratio $\alpha = 0.0182$ (Figure 3-19). The parameter η in Wen-Bouc (Wen 1976) model defines the shape of the hysteresis loop. The chosen values of η and α attempt to match the stiffness and damping characteristics of the building in static and dynamic analyses. The building is subjected to the east and north direction ground motions shown in Figure 1-1. Due to the flexibility of columns and braces, the inter-story drift ratio is not the same in every story. The maximum drift ratio is found to occur in the tenth story. Figure 4-15 shows histories of total (elastic plus plastic) drift ratios in the east and north directions. It is likely that the drift ratios can increase by changing the orientation of the building with respect to the ground motion. Therefore, the rotations in two orthogonal directions are combined to generate the resultant rotations shown in Figure 4-16. The maximum drift ratio is 0.0163 radian, and it occurs in the direction that is 89 degrees clockwise from north (N89E). Figure 4-17 shows the history of drift ratio in the N89E direction.

The equivalent number of cycles is determined as described in Section 1.11. Each "zero-crossing" half-cycle is assigned a weight that is proportional to the square of its amplitude. Elastic perturbations that do not cross the "zero line" are not counted because they do not cause damage. Adding the weights and dividing by 2 gives the equivalent number of cycles to be $N_{eq} = 1$. Therefore, per dynamic analysis, peak drift ratio of 0.0163 radian occurs for one cycle. Figure 4-18 shows resultant roof

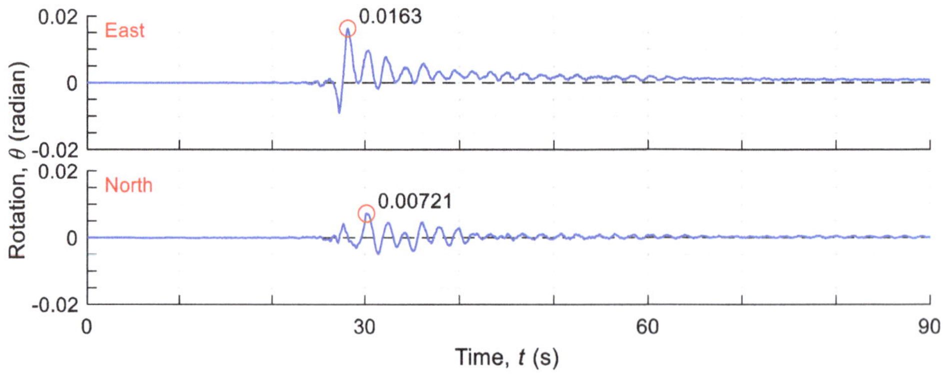

Figure 4-15. Tenth-story drift ratios in 20-story steel eccentrically braced frame building during the TAPS ground motion.

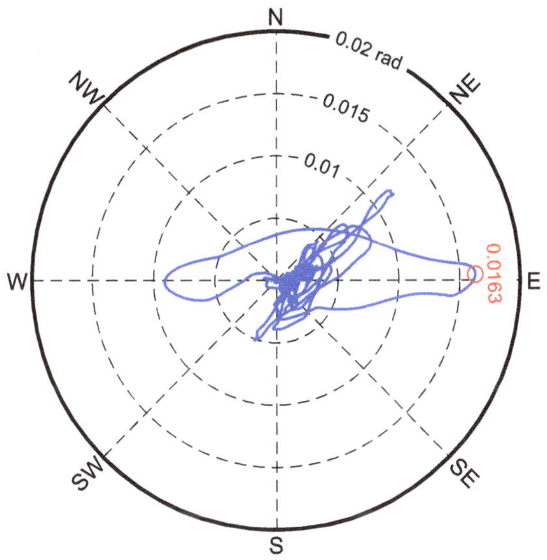

Figure 4-16. Resultant tenth-story drift ratios in 20-story building during the TAPS ground motion.

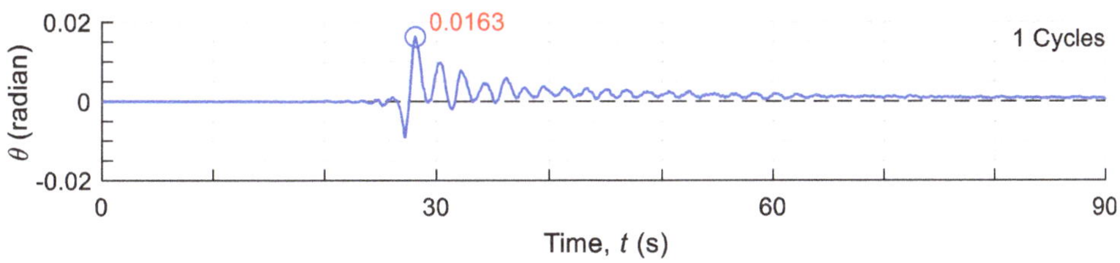

Figure 4-17. Tenth-story drift ratio in N89E direction of 20-story building during the TAPS ground motion.

deflections from dynamic analysis. The maximum deflection is 30.7 in. The second row of Table 4-2 lists the responses of the building from the nonlinear-dynamic analysis.

4.9 Damage Suffered by 20-Story Building

Per static analysis, the drift ratio is 0.0183 radian for 2.33 cycles, but the maximum drift capacity is $\theta_{ult} = 0.0546$ radian for one cycle. Per static analysis, the structural damage suffered by the building during the TAPS ground motion is $0.0183^2 \times 2.33/(0.0546^2 \times 1) = 0.26$ (26%). Per dynamic analysis, the story drift ratio is 0.0163 radian for one cycle, but the maximum drift capacity is $\theta_{ult} = 0.0546$ radian for one cycle. Per dynamic analysis, the structural damage suffered by the building during the

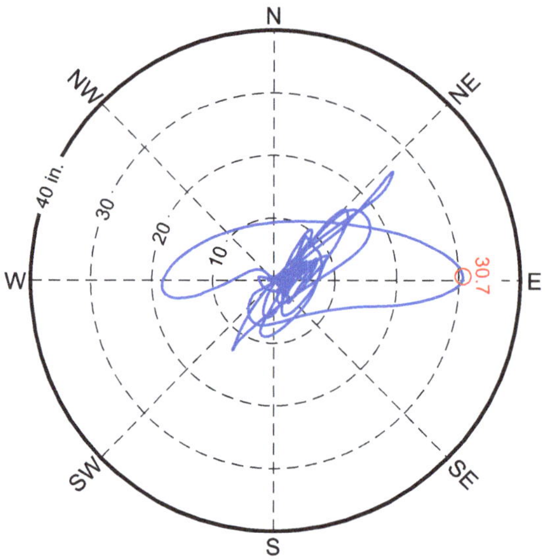

Figure 4-18. Resultant roof deflection of 20-story steel eccentrically braced frame building during the TAPS ground motion.

TAPS ground motion is $0.0163^2 \times 1/(0.0546^2 \times 1) = 0.09$ (9%). From both static and dynamic analyses, the damage suffered by the building during the TAPS ground motion is small. Therefore, the building should be able to withstand additional damage from aftershocks without the danger of collapse.

4.10 Summary

1. The damage suffered by a structure depends on (1) peak deformation, and (2) equivalent number of cycles.
2. There are two methods of determining the peak deformation and equivalent number of cycles: (1) nonlinear-static analysis, and (2) nonlinear dynamic analysis.
3. Nonlinear-static analysis is approximate, but it is transparent and free of numerical errors. Nonlinear-dynamic analysis can be more "accurate", but it is somewhat opaque and can be corrupted by numerical errors.
4. Nonlinear-static analysis should always precede a nonlinear-dynamic analysis.
5. The deformation of a structure generally reduces with increases in strength and damping.
6. The cyclic demand on a structure reduces with increase in effective period and damping. For a ductile-hysteretic structure both period and damping increase with deformation. Therefore, cyclic demand reduces with increase in deformation.

7. A structure cannot be considered deformable if it lacks clearance all around to deform freely.

References

AISC (American Institute of Steel Construction). 2016. *Seismic provisions for structural steel buildings.* AISC 341-16. Chicago: AISC.

ASCE. 2017. *Seismic evaluation and retrofit of existing buildings.* ASCE 41-17. Reston, VA: ASCE.

CSI (Computers and Structures, Inc.). 2011. *SAP 2000.* Berkeley, CA: CSI.

Malhotra, P. K. 2021. *Seismic analysis of structures and equipment.* Cham, Switzerland: Springer.

Wen, Y. K. 1976. "Method for random vibration of hysteretic systems." *J. Eng. Mech. Div.* 102 (2): 249–263.

CHAPTER 5

Collapse-Inducing Ground Motions

5.1 Introduction

In this chapter, the weakest ground motions that can collapse the 3-story steel SMF and 20-story steel EBF buildings are determined. In the next chapter, the responses of 3- and 20-story buildings to the strongest recorded ground motions are determined. For the sake of simplicity, cyclic demand is not considered in this chapter, but it will be considered in the next chapter.

5.2 Weakest Collapse-Inducing Ground Motion

Because energy demand, *ED*, is proposed as a measure of strength of ground shaking, the weakest ground motions are those with the lowest energy demand. The following steps are used to determine the weakest ground motion that can collapse the 3-story steel SMF building:

1. Assume a certain value of PGA based on judgment.
2. Assume normalized velocity $PGV_n = 1$ and central period $T_c = 1.2T_{ult}$. This is because (1) for the same energy, narrow-banded ground motions with energy centroid matching the ultimate period of the building are found to be most destructive, and (2) the centroid of damping-adjusted ADRS (demand curve) is roughly 1/1.2 times the centroid of 5% damped ADRS.
3. Using the definitions of T_c [Equation (1-3)] and PGV_n [Equation (1-4)], calculate $PGD = PGA\left(\frac{T_c}{2\pi}\right)^2$, and $PGV = PGV_n \cdot (PGA \times PGD)^{1/2}$.
4. From PGA, PGV, and PGD, generate smooth ADRS for various values of damping per Malhotra (2021b).
5. Using the damping curve for the 3-story building, and the ADRS for various values of damping, generate damping-adjusted demand curve as described in Section 4.2.
6. Calculate energy centroid T_{cg} from the damping-adjusted demand curve. If $T_{cg} \neq T_{ult}$, multiply T_c by T_{ult}/T_{cg} and restart at Step 3.

7. Superimpose the capacity curve of the building (Figure 3-7) on the demand curve.

   If the demand curve intersects the capacity curve, the building does not collapse. Increase PGA and restart at Step 1.

   If the demand curve envelopes the capacity curve without intersecting it, the building collapses. Decrease PGA and restart at Step 1.
8. Determine PGA, PGV, and PGD that just collapse the building. Determine ED from the 5% damping ADRS generated from PGA, PGV, and PGD.

The PGA, PGV, and PGD of the weakest ground motion that collapses the 3-story steel SMF building are $PGA = 0.594g$, $PGV = 75.9$ in./s, and $PGD = 25.1$ in. The 5% damping response spectrum of this ground motion is shown in Figure 5-1. The energy demand for this ground motion is $ED = 23.8\ (\text{m/s})^2$.

Figure 5-2 shows the ADRS of collapse-inducing ground motion for various values of damping. Figure 5-3 shows the damping-adjusted demand curve for the 3-story building generated per Section 4.2. Notice that the energy centroid $T_{cg} = 1.79$ s matches the ultimate period of the 3-story building T_{ult}. Figure 5-4 shows the demand-capacity equilibrium and energy diagram for the building.

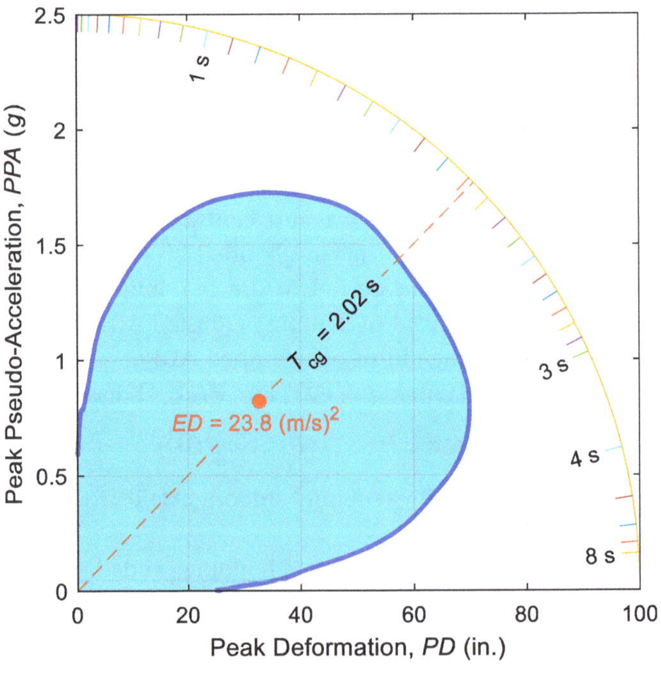

Figure 5-1. Five-percent damping acceleration-deformation response spectrum of weakest ground motion that collapses the 3-story steel special moment frame building as described in Chapter 3.

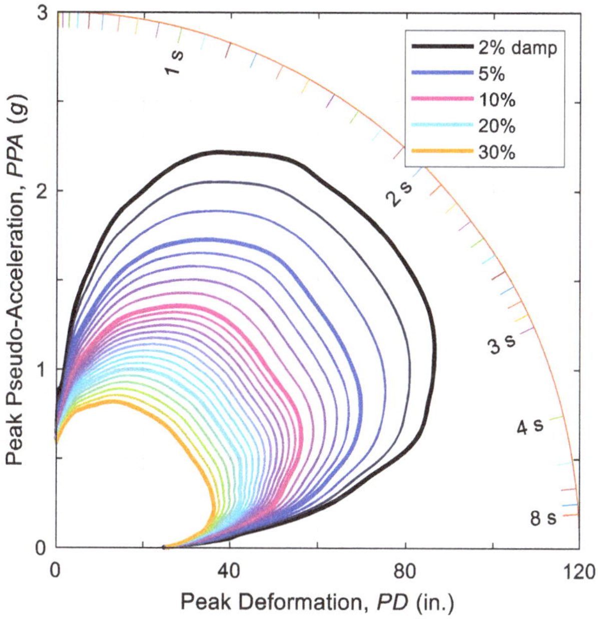

Figure 5-2. Various damping acceleration-deformation response spectra of weakest ground motion that collapses the 3-story steel special moment frame building.

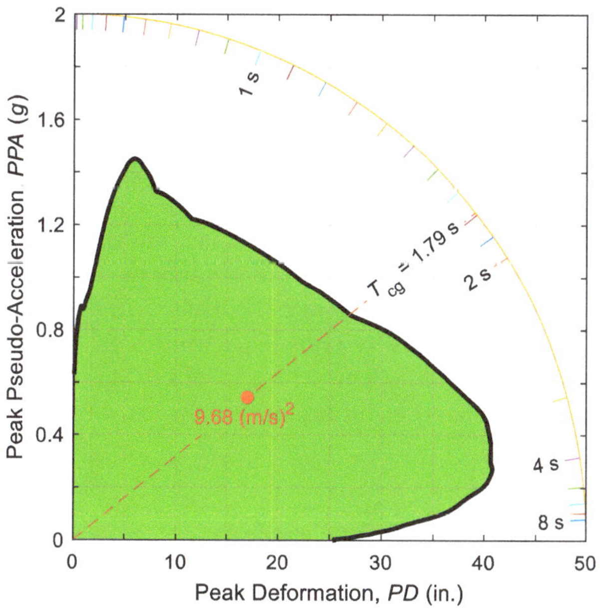

Figure 5-3. Damping-adjusted demand curve of weakest ground motion that collapses the 3-story steel special moment frame building.

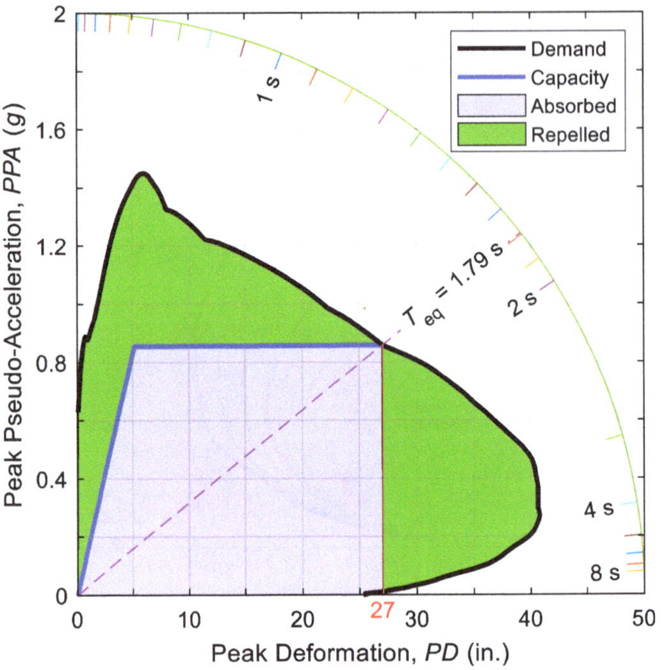

Figure 5-4. Demand-capacity equilibrium and energy diagram for the 3-story steel special moment frame building subjected to the weakest collapse-inducing ground motion.

The building is barely collapsed by this ground motion. The first row of Table 5-1 lists the parameters of weakest ground motion that can collapse the 3-story steel SMF building.

To confirm that the results from nonlinear-static analysis are conservative, a collapse-inducing ground motion history is generated by matching its response spectrum with the collapse-inducing response spectrum shown in Figure 5-1. The method described in Malhotra (2021a) is used to generate the spectrum-compatible ground motion history. Figure 5-5 shows the collapse-inducing ground motion history. Figure 5-6 compares its response spectrum with the collapse-inducing response spectrum shown in Figure 5-1. The *PGA*, *PGV*, and *PGD* of collapse-inducing ground motion history are somewhat different from the *PGA*,

Table 5-1. Weakest Ground Motions That Collapse the 3-Story Steel SMF Building and the 20-Story Steel Eccentrically Braced Frame Building.

Building	*PGA* (*g*)	*PGV* (in./s)	*PGD* (in.)	*ED* (m/s)^2
3-story steel SMF	0.594	75.9	25.1	23.8
20-story steel EBF	0.278	90.9	77.0	34.1

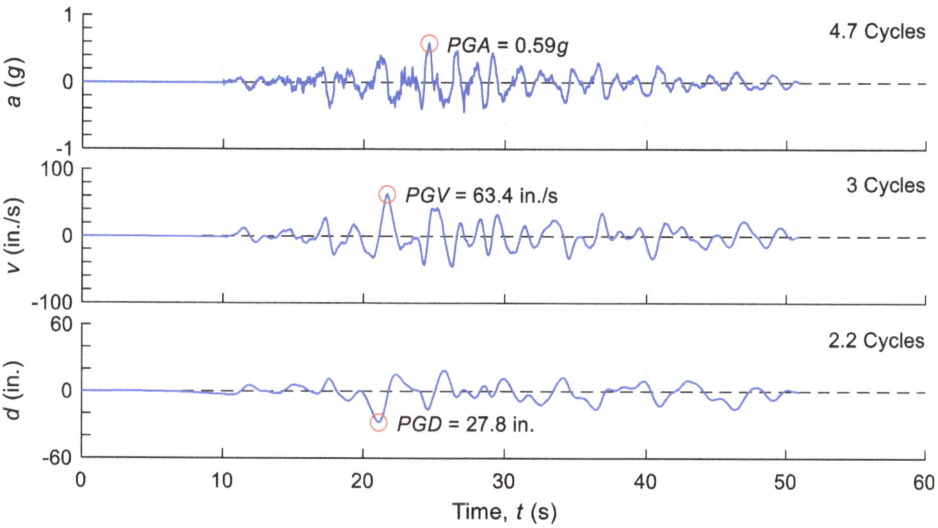

Figure 5-5. Ground motion history that is compatible with the collapse-inducing response spectrum shown in Figure 5-1.

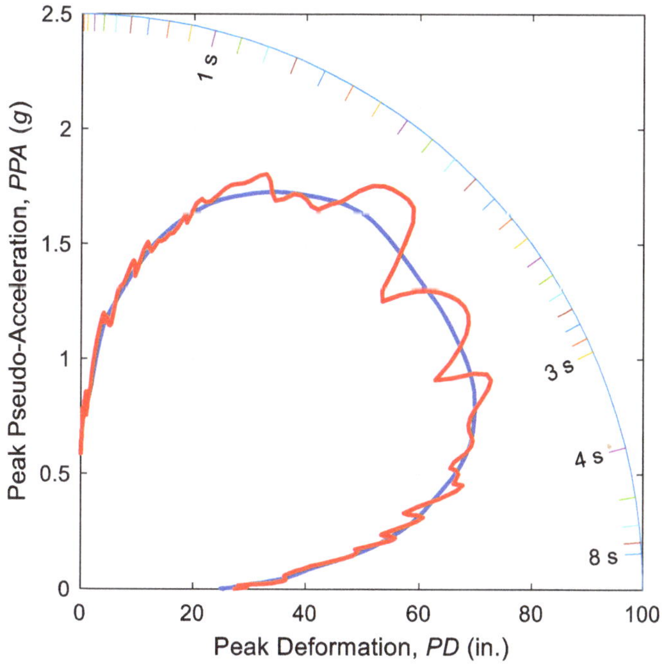

Figure 5-6. Response spectrum of ground motion shown in Figure 5-5 compared with the collapse-inducing response spectrum shown in Figure 5-1.

PGV, and PGD used to generate the collapse-inducing response spectrum. This difference is because of the "average" shape of the smooth response spectrum generated from PGA, PGV, and PGD. Notice that the dominant frequencies in acceleration, velocity and displacement histories are similar. This is the nature of a narrow-banded ground motion that is dominated by a single frequency.

The response of the 3-story steel SMF building is calculated from dynamic analysis as described in Chapter 4. The maximum inter-story drift of 0.0646 radian is found to occur in the first story. This is less than the ultimate inter-story drift of $\theta_{ult} = 0.104$ radian. Therefore, the collapse-inducing ground motions obtained from the nonlinear-static analysis are conservative.

Next, the 5% damping ADRS of collapse-inducing ground motion is determined for the 20-story steel EBF building. It is shown in Figure 5-7. The energy demand for this ground motion is $ED = 34.1\ (\mathrm{m/s})^2$. Figure 5-8 shows the ADRS of collapse-inducing ground motion for various values of damping. Figure 5-9 shows the damping-adjusted demand curve for the 20-story building generated per Section 4.2. Once again, the energy centroid $T_{cg} = 4.7\,\mathrm{s}$ matches the ultimate period of the building T_{ult}. Figure 5-10 shows the demand-capacity equilibrium and energy diagram for the 20-story building. The building is barely collapsed by this ground

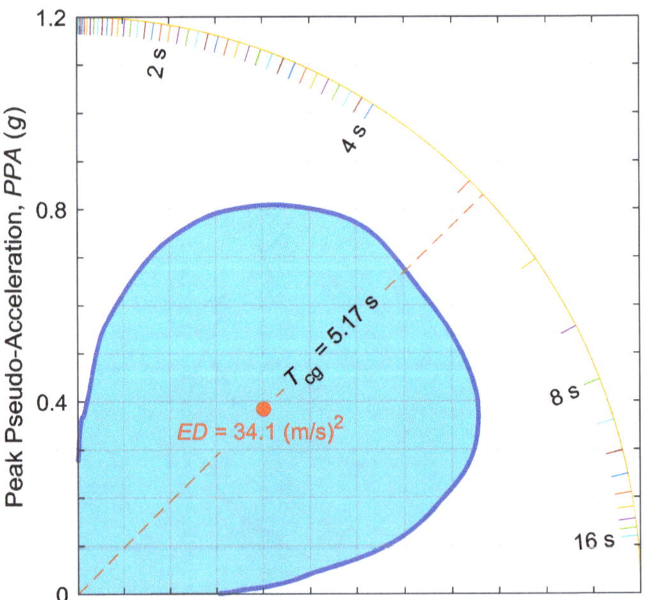

Figure 5-7. Five-percent damping acceleration-deformation response spectrum of weakest ground motion that collapses the 20-story steel eccentrically braced frame building.

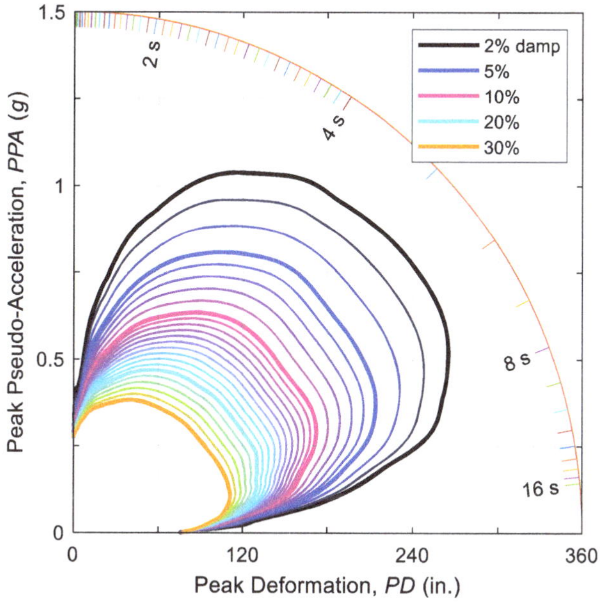

Figure 5-8. Various damping acceleration-deformation response spectra of weakest ground motion that collapses the 20-story steel eccentrically braced frame building.

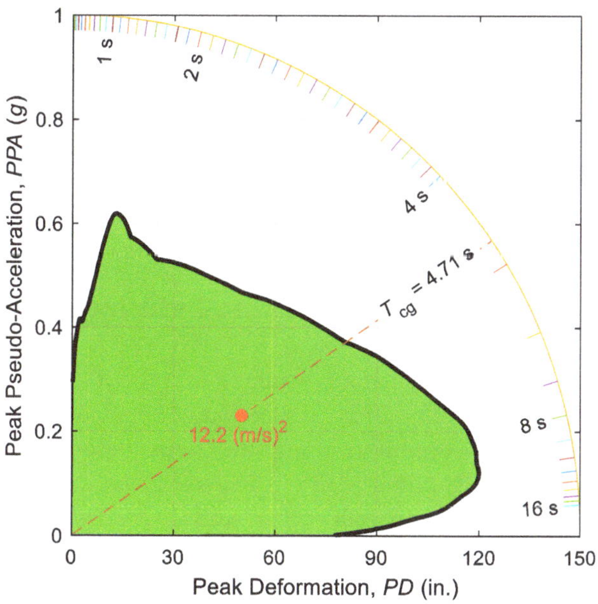

Figure 5-9. Damping-adjusted demand curve of weakest ground motion that collapses the 20-story steel eccentrically braced frame building.

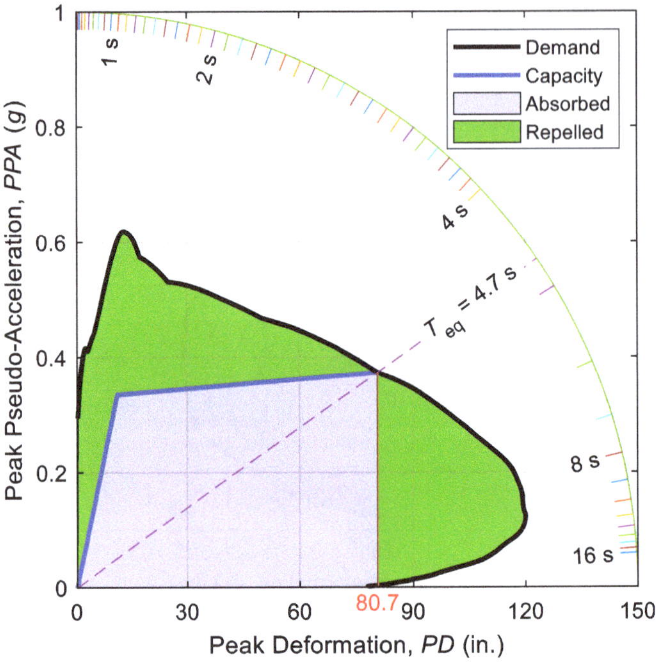

Figure 5-10. Demand-capacity equilibrium and energy diagram for the 20-story steel eccentrically braced frame building subjected to the weakest collapse-inducing ground motion.

motion. The second row of Table 5-1 lists the parameters of the weakest ground motion that collapses the 20-story steel EBF building. Higher energy demand needed to collapse the 20-story building is due to higher seismic toughness ST and higher damping for this building.

5.3 Effect of Deformability on Collapse-Inducing Ground Motion

To study the effect of deformability, it is assumed that the 3-story building collapses at 15 in. deflection rather than 27 in. deflection. The weakest collapse-inducing ground motion is again computed using the steps described in Section 5.2. Figure 5-11 shows the 5% damping ADRS of weakest ground motions that collapse 3-story steel buildings of deformability 27 in. and 15 in. Table 5-2 compares the parameters of collapse-inducing ground motions for two values of deformability. A 44% reduction in deformability of 3-story building causes 22% reduction in PGA, 42% reduction in PGV, 57% reduction in PGD, and 67% reduction in ED of collapse-inducing ground motion. This underscores the role of deformability in preventing collapse.

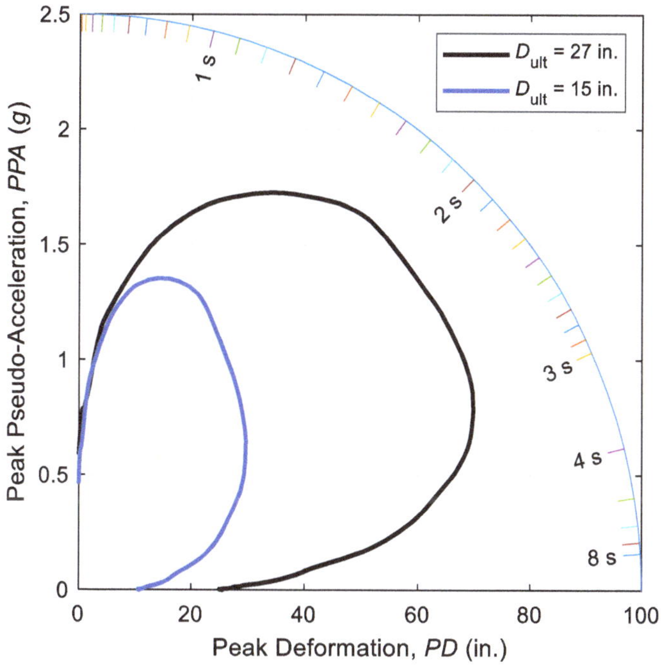

Figure 5-11. Five-percent damping acceleration-deformation response spectra of weakest ground motions that collapse 3-story steel buildings of deformability 27 and 15 in.

Table 5-2. Effect of Deformability on Weakest Ground Motion That Collapses the 3-Story Building.

D_{ult} (in.)	PGA (g)	PGV (in./s)	PGD (in.)	ED $(m/s)^2$
27	0.594	75.9	25.1	23.8
15	0.466	43.8	10.7	7.9

5.4 Effect of Hysteresis on Collapse-Inducing Ground Motion

As seen in Chapter 3, hysteresis plays an important role in increasing the damping of the building from elastic damping of 2% to more than 20% of critical at collapse. To study the effect of hysteresis, it is assumed that the 3-story building collapses at 27 in. deflection, but its damping does not increase due to hysteresis; it remains at 2% of critical. In other words, the building is ductile but non-hysteretic. This can happen if yielding occurs in connections rather than in beams. The weakest collapse-inducing ground motion is computed using the steps described in Section 5.2. Figure 5-12 shows the 5% damping ADRS of weakest ground motions that collapse 3-story steel

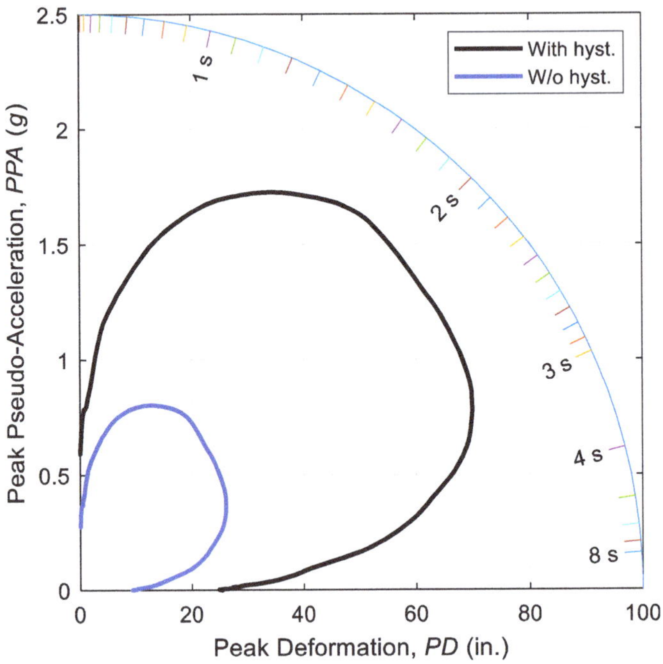

Figure 5-12. Five-percent damping acceleration-deformation response spectra of weakest ground motions that collapse the 3-story steel building with and without hysteretic damping.

building with and without hysteretic damping. Table 5-3 compares the parameters of collapse-inducing ground motions with and without hysteretic damping. Lack of hysteresis causes 54% reduction in PGA, 58% reduction in PGV, 62% reduction in PGD, and 82% reduction in ED of collapse-inducing ground motion. This underscores the importance of hysteresis in preventing collapse.

Deformability and hysteresis are very important in collapse prevention, but in prescriptive seismic design, deformability and hysteresis are not explicitly considered. Their effect is considered by applying judgment-based load reduction R factors. This makes the prescriptive seismic design quite arbitrary.

Table 5-3. Effect of Hysteresis on Weakest Ground Motion That Collapses the 3-Story Building.

Hysteresis	$PGA\,(g)$	PGV (in./s)	PGD (in.)	$ED\,(\text{m/s})^2$
Yes	0.594	75.9	25.1	23.8
No	0.276	31.7	9.44	4.2

5.5 Summary

1. The weakest ground motion that collapses the 3-story steel SMF building described in Chapter 3 has an energy-demand of $ED = 23.8\,(\text{m/s})^2$. The peak ground acceleration, velocity, and displacement for this ground motion are $PGA = 0.594g$, $PGV = 75.9\,(\text{in./s})$, and $PGD = 25.1$ in. Any other ground motion that collapses this building has $ED > 23.8\,(\text{m/s})^2$.
2. The weakest ground motion that collapses the 20-story steel EBF building described in Chapter 3 has an energy demand of $ED = 34.1\,(\text{m/s})^2$. The peak ground acceleration, velocity, and displacement for this ground motion are $PGA = 0.278g$, $PGV = 90.9$ in./s, and $PGD = 77$ in. Any other ground motion that collapses this building has $ED > 34.1\,(\text{m/s})^2$.
3. The higher energy demand ED needed to collapse the 20-story building is because of higher seismic toughness ST and slightly higher damping for that building as discussed in Chapter 3.
4. The 20-story building needs smaller ground accelerations to yield but larger ground velocities and displacements to collapse than the 3-story building.
5. Narrow-banded ground motions with energy centroid T_{cg} matching the ultimate period T_{ult} of the building are most destructive.
6. In collapse prevention, deformability and hysteresis are as important as strength, hence they should be explicitly considered in seismic design.
7. The most important seismic characteristic of a building is the minimum energy demand ED needed to collapse that building.

References

Malhotra P. K. 2021a. "Ground motions for future earthquakes." Chap. 2 in *Seismic analysis of structures and equipment*. Cham, Switzerland: Springer. https://doi.org/10.1007/978-3-030-57858-9_2.

Malhotra P. K. 2021b. "Ground motions from past earthquakes." Chap. 1 in *Seismic analysis of structures and equipment*. Cham, Switzerland: Springer. https://doi.org/10.1007/978-3-030-57858-9_1.

CHAPTER 6

Response of Buildings to Strongest Ground Motions

6.1 Introduction

In this chapter, the 3- and 20-story buildings described in Chapter 3 are subjected to 11 strongest ground motions listed in Table 1-2. The buildings' responses are calculated by using both static and dynamic methods of nonlinear analysis discussed in Chapter 4. Cumulative fatigue damage from cycles of various amplitudes is considered. The analysis steps are not repeated; only the main results are discussed.

6.2 Northridge, Tarzana Ground Motion

This ground motion was recorded at Tarzana Cedar Hill A station, at 0.37 km from vertical projection of the fault, during the 1994 magnitude **M** 6.7 Northridge, California Earthquake (Darragh et al. 1994). The recording station is owned and maintained by the California Geological Survey (CGS); it was originally situated on top of a hill on Site Class D soil as defined by ASCE 7-22 (ASCE 2022). The California Geological Survey processed the ground motion data by applying baseline correction and low- and high-pass filters. The lowest usable frequency is 0.1 Hz, or the longest usable period is $T_u = 1/0.1 = 10$ s. Figure 6-1 shows acceleration, velocity, and displacement histories of Tarzana ground motion in maximum horizontal (RotD100) directions. Recall from Chapter 1 that RotD100 directions for acceleration, velocity, and displacement are not same. The peak ground acceleration, velocity, and displacement in RotD100 directions are $PGA = 2.01g$, $PGV = 44.2$ in./s, and $PGD = 16.3$ in. The number of cycles in velocity history is 2.2. Unusually high PGA at this station could not be attributed to a single cause (Graizer and Shakal 2003). With energy demand $ED = 10.3$ (m/s)^2, energy centroid $T_{cg} = 0.574$ s, and normalized velocity $PGV_n = 0.39$, Tarzana ground motion is extremely strong, medium- to high-frequency, and broad banded.

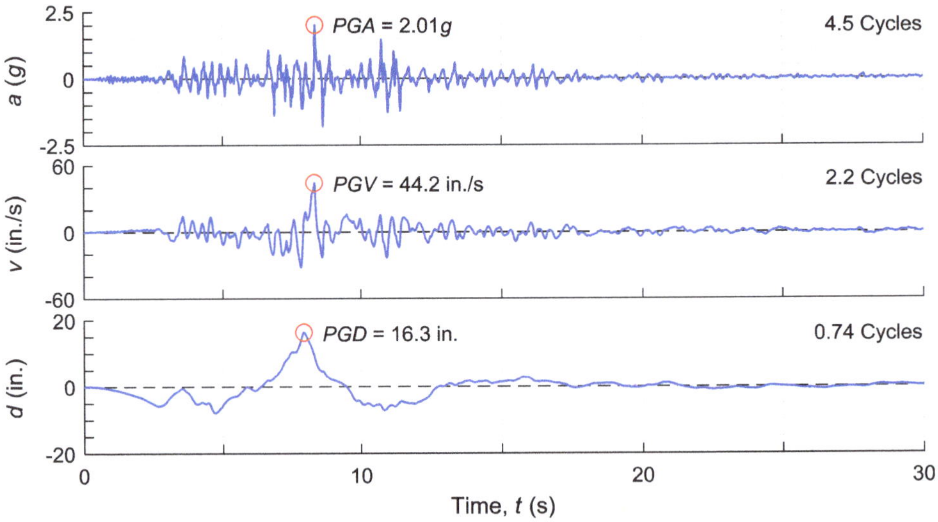

Figure 6-1. Acceleration, velocity, and displacement histories of Tarzana ground motion in maximum-horizontal (RotD100) directions.

6.2.1 Response of 3-Story Building to Tarzana Ground Motion

Per Table 5-1, the minimum *ED* required to collapse the 3-story steel SMF building is 23.8 $(m/s)^2$. The *ED* of Tarzana ground motion is only 10.3 $(m/s)^2$ and it is centered away from the ultimate period of the building ($T_{cg} = 0.32\ T_{ult}$). Cyclic demand is moderate. Therefore, Tarzana ground motion cannot collapse the 3-story building, but it can damage the building. The response of the building is calculated from both static and dynamic methods of nonlinear analysis described in Chapter 4.

Figure 6-2 shows the demand-capacity equilibrium for the building subjected to Tarzana ground motion. Two demand curves are shown: (1) The magenta curve is the hypothetical demand-curve without any hysteretic damping; it is the 2% damping ADRS; and (2) the black curve is the actual demand-curve for the building generated by the procedure described in Section 4.2. Per static analysis, the CG of the building deflects 8.1 in. The effective period of the building at equilibrium is $T_{eq} = 0.99$ s, and effective damping at equilibrium is $\zeta_{eq} = 5.2\%$. Using T_{eq} and ζ_{eq}, the equivalent number of cycles, read from cyclic-demand spectra (not shown), is $N_{eq} = 2.6$. Therefore, the CG of the building deflects 8.1 in. for 2.6 cycles. It was established in Chapter 3 that the CG of the building has the capacity to deflect 27 in. for two cycles. Therefore, fatigue damage suffered by the building is $(8.1^2 \times 2.6)/(27^2 \times 2) = 0.12$ (12%) per static analysis. Notice in Figure 6-2 that the high-frequency energy, associated with high *PGA*, is not absorbed by the building. Table 6-1 provides a summary of results from static and dynamic analyses. The results from static analysis are more conservative in this case. If the building had no

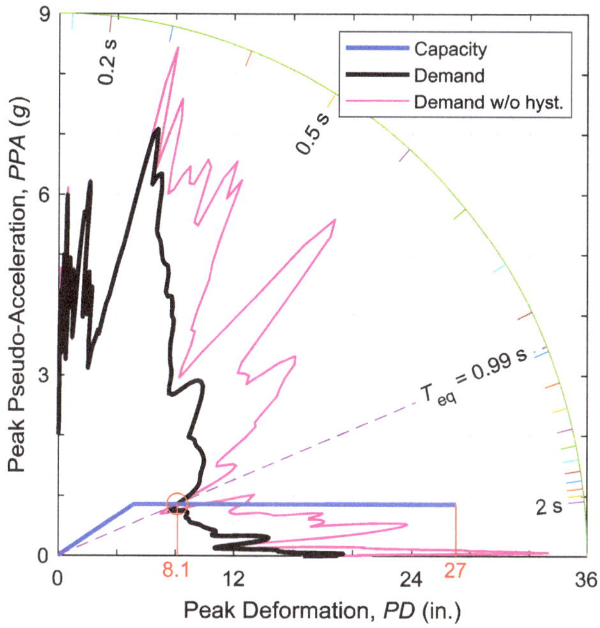

Figure 6-2. Demand-capacity equilibrium for 3-story building subjected to the Tarzana ground motion shown in Figure 6-1.

Table 6-1. Response of 3-Story Building to Tarzana Ground Motion.

Nonlinear analysis	Roof deflection (in.)	Maximum drift ratio (radian)	Number of cycles	Fatigue damage (%)
Static	13.6	0.0314	2.6	12
Dynamic	10.4	0.0299	1.4	6

hysteretic damping (magenta curve in Figure 6-2), it would deflect 8.1 in. for 4.6 cycles and suffer 20% fatigue damage.

6.2.2 Response of 20-Story Building to Tarzana Ground Motion

Per Table 5-1, the minimum *ED* required to collapse the 20-story steel EBF building is 34.1 $(m/s)^2$. The *ED* of Tarzana ground motion is only 10.3 $(m/s)^2$ and it is centered away from the ultimate period of the building ($T_{cg} = 0.12\ T_{ult}$). Cyclic demand is moderate. Therefore, Tarzana ground motion cannot collapse the 20-story building, but it can damage the building. The response of the building is calculated from both static and dynamic methods of nonlinear analysis described in Chapter 4.

Figure 6-3 shows the demand-capacity equilibrium for the building subjected to Tarzana ground motion. Two demand curves are shown: (1) The magenta curve is

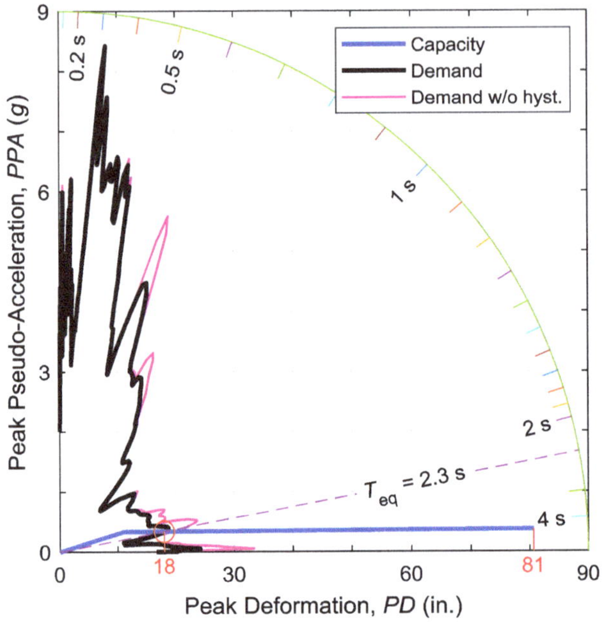

Figure 6-3. Demand-capacity equilibrium for 20-story building subjected to Tarzana ground motion shown in Figure 6-1.

the hypothetical demand curve without any hysteretic damping; it is the 2% damping ADRS; and (2) the black curve is the actual demand curve for the building generated by the procedure described in Section 4.2. Per static analysis, the CG of the building deflects 18 in. The effective period of the building at equilibrium is $T_{eq} = 2.3$ s, and effective damping at equilibrium is $\zeta_{eq} = 5.5\%$. Using T_{eq} and ζ_{eq}, the equivalent number of cycles, read from cyclic-demand spectra (not shown), is $N_{eq} = 1.1$. Therefore, the CG of the building deflects 18 in. for 1.1 cycles. It was established in Chapter 3 that the CG of the building has the capacity to deflect 81 in. for one cycle. Therefore, fatigue damage suffered by the building is $(18^2 \times 1.1)/(81^2 \times 1) = 0.10$ (10%). Notice in Figure 6-3 that the high-frequency energy, associated with high *PGA*, is not absorbed by the building. Table 6-2 provides a

Table 6-2. Response of 20-Story Building to Tarzana Ground Motion.

Nonlinear analysis	Roof deflection (in.)	Maximum drift ratio (radian)	Number of cycles	Fatigue damage (%)
Static	35	0.0121	2.1	10
Dynamic	19.7	0.0128	1.1	6

summary of results from static and dynamic analyses. The results from static analysis are more conservative in this case. If the building had no hysteretic damping (magenta curve in Figure 6-3), it would deflect 19.6 in. for 6.2 cycles and suffer 37% fatigue damage.

6.3 Loma Prieta, Los Gatos Presentation Center Ground Motion

This ground motion was recorded at the LGPC (Los Gatos Presentation Center) station, at zero distance from vertical projection of the fault, during the 1989 magnitude **M** 6.9 Loma Prieta, California Earthquake (PEER 2022). The recording station is owned and maintained by the University of California, Santa Cruz (UCSC). It is situated on Site Class C soil as defined by ASCE 7-22. The ground motion was processed by applying baseline correction and low- and high-pass filters (PEER 2022). The lowest usable frequency is 0.125 Hz, or the longest usable period is $T_u = 1/0.125 = 8$ s. Figure 6-4 shows acceleration, velocity, and displacement histories of LGPC ground motion in maximum-horizontal (RotD100) directions. The peak ground acceleration, velocity, and displacement in RotD100 directions are $PGA = 1.03g$, $PGV = 43$ in./s, and $PGD = 26.8$ in. The number of cycles in velocity history is 2. With energy demand $ED = 11.8\ (\text{m/s})^2$, energy centroid $T_{cg} = 1.64$ s, and normalized velocity $PGV_n = 0.42$, LGPC ground motion is extremely strong, medium to low frequency, and broad banded.

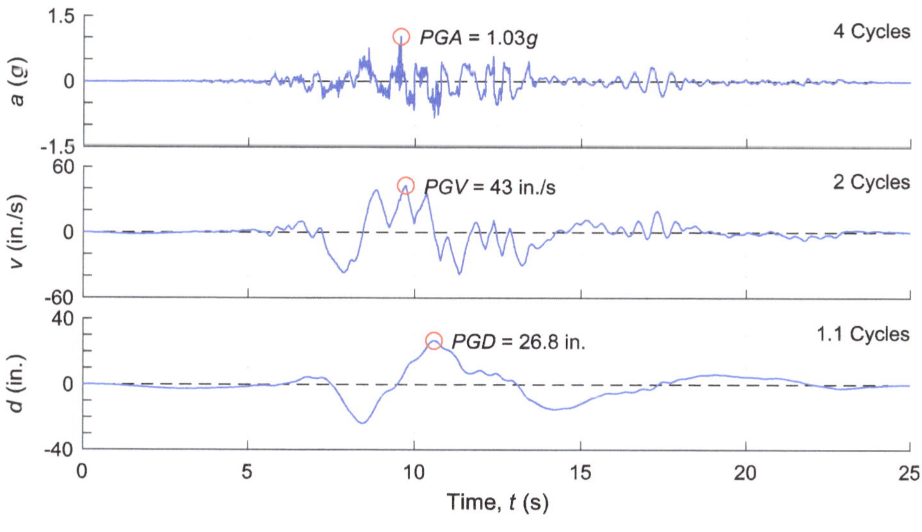

Figure 6-4. Acceleration, velocity, and displacement histories of Los Gatos Presentation Center ground motion in maximum-horizontal (RotD100) directions.

6.3.1 Response of 3-Story Building to Los Gatos Presentation Center Ground Motion

Per Table 5-1, the minimum *ED* required to collapse the 3-story steel SMF building is 23.8 $(m/s)^2$. The *ED* of LGPC ground motion is only 11.8 $(m/s)^2$. It is centered close to the ultimate period of the building ($T_{cg} = 0.92\ T_{ult}$), but it is not narrowly focused ($PGV_n = 0.42$). Cyclic demand is moderate. Therefore, LGPC ground motion cannot collapse the 3-story building, but it can damage the building. The response of the building is calculated from both static and dynamic methods of nonlinear analysis described in Chapter 4.

Figure 6-5 shows the demand–capacity equilibrium for the building subjected to LGPC ground motion. Two demand curves are shown: (1) The magenta curve is the hypothetical demand curve without any hysteretic damping; it is the 2% damping ADRS; and (2) the black curve is the actual demand curve for the building generated by the procedure described in Section 4.2. Per static analysis, the CG of the building deflects 10 in. The effective period of the building at equilibrium is $T_{eq} = 1.1$ s, and effective damping at equilibrium is $\zeta_{eq} = 8.2\%$. Using T_{eq} and ζ_{eq}, the equivalent number of cycles, read from cyclic-demand spectra (not shown), is $N_{eq} = 2.4$. Therefore, the CG of the building deflects 10 in. for 2.4 cycles. It was established in Chapter 3 that the CG of the building has the capacity to deflect 27 in. for two

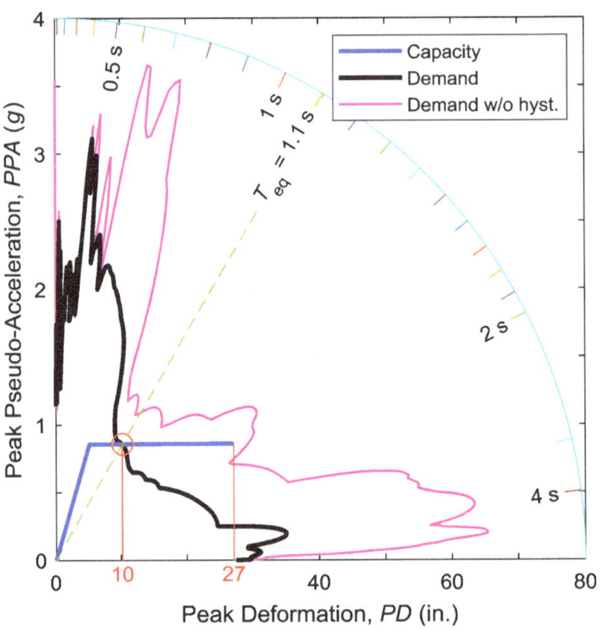

Figure 6-5. Demand-capacity equilibrium for 3-story building subjected to LGPC ground motion shown in Figure 6-4.

Table 6-3. Response of 3-Story Building to Los Gatos Presentation Center Ground Motion.

Nonlinear analysis	Roof deflection (in.)	Maximum drift ratio (radian)	Number of cycles	Fatigue damage (%)
Static	17	0.0394	2.4	17
Dynamic	10.5	0.0301	2.3	10

cycles. Therefore, the fatigue damage suffered by the building is $(10^2 \times 2.4)/(27^2 \times 2) = 0.17$ (17%) per the static analysis. Notice in Figure 6-5 that the high-frequency energy, associated with high *PGA*, is not absorbed by the building. Table 6-3 provides a summary of results from static and dynamic analyses. The results from static analysis are more conservative in this case. If the building had no hysteretic damping (magenta curve in Figure 6-5), it would collapse because the capacity curve stops short of the magenta demand-curve.

6.3.2 Response of 20-Story Building to Los Gatos Presentation Center Ground Motion

Per Table 5-1, the minimum *ED* required to collapse the 20-story steel EBF building is 34.1 $(m/s)^2$. The *ED* of LGPC ground motion is only 10.3 $(m/s)^2$ and it is centered away from the ultimate period of the building ($T_{cg} = 0.35\ T_{ult}$). Cyclic demand is moderate. Therefore, LGPC ground motion cannot collapse the 20-story building, but it can damage the building. The response of the building is calculated from both static and dynamic methods of nonlinear analysis described in Chapter 4.

Figure 6-6 shows the demand-capacity equilibrium for the building subjected to LGPC ground motion. Two demand curves are shown: (1) The magenta curve is the hypothetical demand-curve without any hysteretic damping; it is the 2% damping ADRS; and (2) the black curve is the actual demand curve for the building generated by the procedure described in Section 4.2. Per static analysis, the CG of the building deflects 30 in. The effective period of the building at equilibrium is $T_{eq} = 3$ s, and effective damping at equilibrium is $\zeta_{eq} = 13\%$. Using T_{eq} and ζ_{eq}, the equivalent number of cycles, read from cyclic-demand spectra (not shown), is $N_{eq} = 2.4$. Therefore, the CG of the building deflects 30 in. for 2.4 cycles. It was established in Chapter 3 that the CG of the building has the capacity to deflect 81 in. for one cycle. Therefore, the fatigue damage suffered by the building is $(30^2 \times 2.4)/(81^2 \times 1) = 0.32$ (32%). Notice in Figure 6-6 that the high-frequency energy, associated with high *PGA*, is not absorbed by the building. Table 6-4 provides a summary of results from static and dynamic analyses. The results from static analysis are more conservative in this case. If the building had no hysteretic damping (magenta curve in Figure 6-6), it would deflect 63 in. for two cycles and suffer 100% fatigue damage, resulting in collapse.

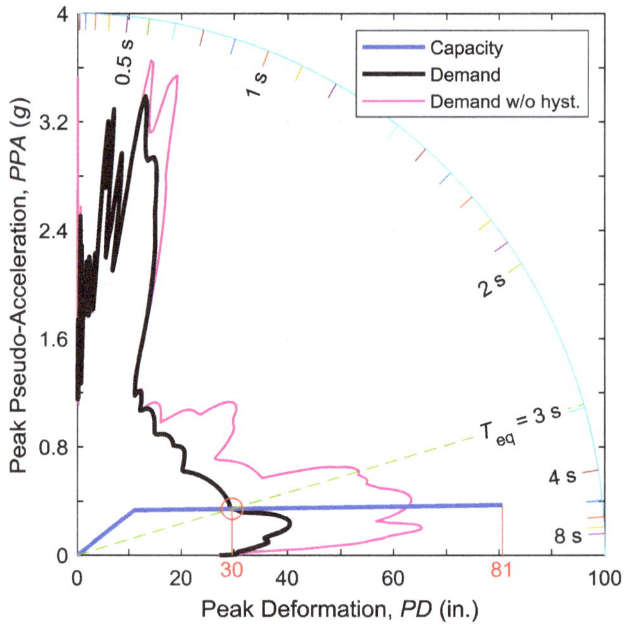

Figure 6-6. Demand-capacity equilibrium for 20-story building subjected to Los Gatos Presentation Center ground motion shown in Figure 6-4.

Table 6-4. Response of 20-Story Building to Los Gatos Presentation Center Ground Motion.

Nonlinear analysis	Roof deflection (in.)	Maximum drift ratio (radian)	Number of cycles	Fatigue damage (%)
Static	58	0.0201	2.4	32
Dynamic	38	0.0192	0.7	9

6.4 Chi-Chi, Taiwan, TCU065 Ground Motion

This ground motion was recorded at the TCU065 station, at 0.59 km from vertical projection of the fault, during the 1999 magnitude **M** 7.6 Chi-Chi, Taiwan Earthquake. The TCU065 station is owned and maintained by the Central Weather Bureau of Taiwan. Per PEER (2022), the TCU065 station is situated on Site Class CD soil as defined by ASCE 7-22 (ASCE 2022). The ground motion was processed by applying baseline correction and low- and high-pass filters (PEER 2022). The lowest usable frequency is 0.075 Hz, or the longest usable period is $T_u = 1/0.075 = 13.3$ s. Figure 6-7 shows acceleration, velocity, and displacement histories of TCU065 ground motion in maximum-horizontal (RotD100) directions. The peak ground acceleration, velocity, and displacement in RotD100 directions are $PGA = 0.85g$, $PGV = 52.6$ in./s, and $PGD = 37.3$ in. The number of cycles in velocity

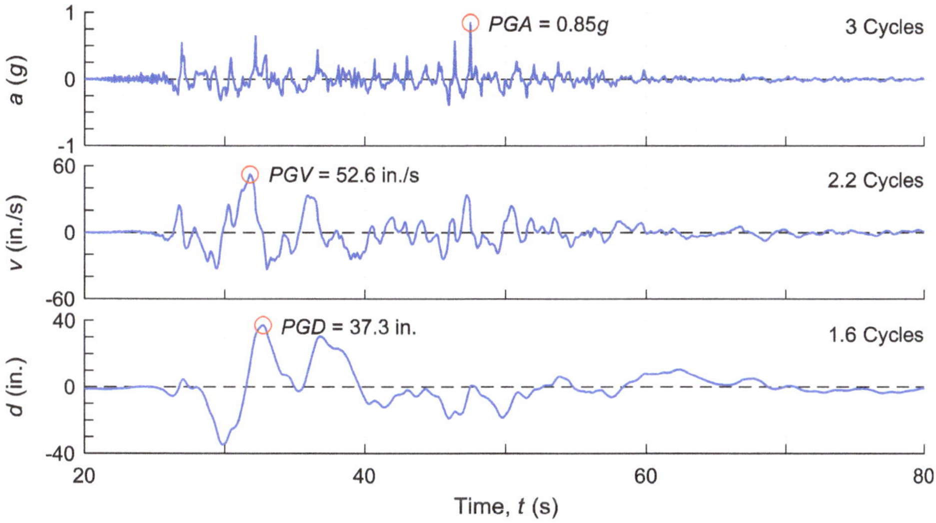

Figure 6-7. Acceleration, velocity, and displacement histories of TCU065 ground motion in maximum-horizontal (RotD100) directions.

history is 2.2. With *ED* = 14.9 $(m/s)^2$, $T_{cg} = 2.72$ s, and $PGV_n = 0.48$, TCU065 ground motion is extremely strong, low-frequency, and medium to broad banded.

6.4.1 Response of 3-Story Building to TCU065 Ground Motion

Per Table 5-1, the minimum *ED* required to collapse the 3-story steel SMF building is 23.8 $(m/s)^2$. The *ED* of TCU065 ground motion is only 14.9 $(m/s)^2$. It is centered close to the ultimate period of the building ($T_{cg} = 1.52\ T_{ult}$), but it is not narrowly focused ($PGV_n = 0.48$). Cyclic demand is moderate. Therefore, TCU065 ground motion cannot collapse the 3-story building, but it can damage the building. The response of the building is calculated from both static and dynamic methods of nonlinear analysis described in Chapter 4.

Figure 6-8 shows the demand-capacity equilibrium for the building subjected to TCU065 ground motion. Two demand curves are shown: (1) The magenta curve is the hypothetical demand curve without any hysteretic damping; it is the 2% damping ADRS; and (2) the black curve is the actual demand curve for the building generated by the procedure described in Section 4.2. Per static analysis, the CG of the building deflects 12 in. The effective period of the building at equilibrium is $T_{eq} = 1.2$ s, and effective damping at equilibrium is $\zeta_{eq} = 11\%$. Using T_{eq} and ζ_{eq}, the equivalent number of cycles, read from cyclic-demand spectra (not shown), is $N_{eq} = 5.7$. Therefore, the CG of the building deflects 12 in. for 5.7 cycles. It was established in Chapter 3 that the CG of the building has the capacity to deflect 27 in. for two

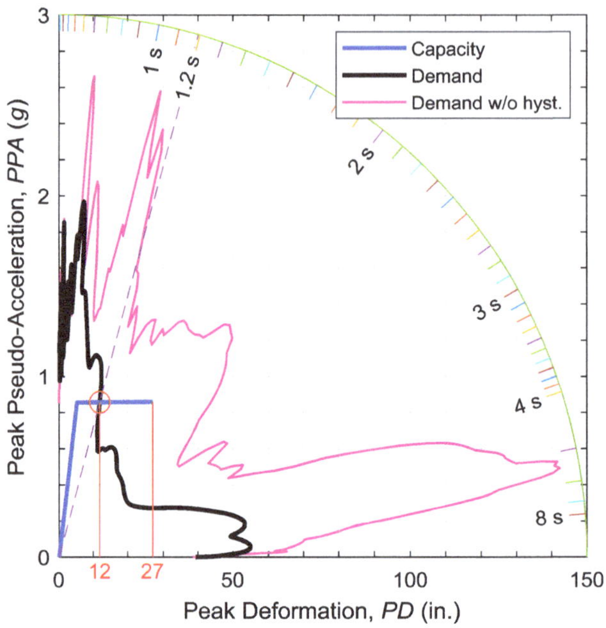

Figure 6-8. Demand-capacity equilibrium and energy diagram for 3-story building subjected to TCU065 ground motion shown in Figure 6-7.

Table 6-5. Response of 3-Story Building to TCU065 Ground Motion.

Nonlinear analysis	Roof deflection (in.)	Maximum drift ratio (radian)	Number of cycles	Fatigue damage (%)
Static	19.8	0.0458	5.7	55
Dynamic	11.2	0.0336	4.1	22

cycles. Therefore, the fatigue damage suffered by the building is $(12^2 \times 5.7)/(27^2 \times 2) = 0.55$ (55%) per the static analysis. Table 6-5 provides a summary of results from static and dynamic analyses. The results from static analysis are more conservative in this case. If the building had no hysteretic damping (magenta curve in Figure 6-8), it would collapse because the capacity curve stops short of the magenta demand curve.

6.4.2 Response of 20-Story Building to TCU065 Ground Motion

Per Table 5-1, the minimum *ED* required to collapse the 20-story steel EBF building is 34.1 $(\mathrm{m/s})^2$. The *ED* of TCU065 ground motion is only 14.9 $(\mathrm{m/s})^2$, and it is centered away from the ultimate period of the building ($T_{cg} = 0.58\ T_{ult}$). Cyclic demand is

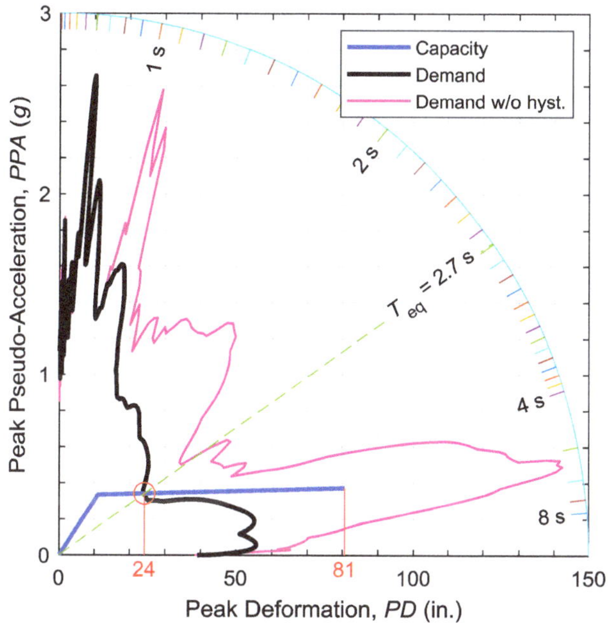

Figure 6-9. Demand-capacity equilibrium for 20-story building subjected to TCU065 ground motion shown in Figure 6-7.

moderate. Therefore, TCU065 ground motion cannot collapse the 20-story building, but it can damage the building. The response of the building is calculated from both static and dynamic methods of nonlinear analysis described in Chapter 4.

Figure 6-9 shows the demand-capacity equilibrium for the building subjected to TCU065 ground motion. Two demand curves are shown: (1) The magenta curve is the hypothetical demand-curve without any hysteretic damping; it is the 2% damping ADRS; and (2) the black curve is the actual demand curve for the building generated by the procedure described in Section 4.2. Per static analysis, the CG of the building deflects 24 in. The effective period of the building at equilibrium is $T_{eq} = 2.7$ s, and effective damping at equilibrium is $\zeta_{eq} = 9.6\%$. Using T_{eq} and ζ_{eq}, the equivalent number of cycles, read from cyclic-demand spectra (not shown), is $N_{eq} = 3.4$. Therefore, the CG of the building deflects 24 in. for 3.4 cycles. It was established in Chapter 3 that the CG of the building has the capacity to deflect 81 in. for one cycle. Therefore, the fatigue damage suffered by the building is $(24^2 \times 3.4)/(81^2 \times 1) = 0.31$ (31%). Notice in Figure 6-9 that the high-frequency energy, associated with high *PGA*, is not absorbed by the building. Table 6-6 provides a summary of results from static and dynamic analyses. The results from static analysis are more conservative in this case. If the building had no hysteretic damping (magenta curve in Figure 6-9), it would collapse because the capacity curve stops short of the magenta demand curve.

Table 6-6. Response of 20-Story Building to TCU065 Ground Motion.

Nonlinear analysis	Roof deflection (in.)	Maximum drift ratio (radian)	Number of cycles	Fatigue damage (%)
Static	47.3	0.0164	3.4	31
Dynamic	29.5	0.0146	2.4	17

6.5 Northridge, Sylmar Ground Motion

This ground motion was recorded at Sylmar Converter station, at zero distance from vertical projection of the fault, during the 1994 magnitude **M** 6.7 Northridge, California Earthquake. The recording station is owned and maintained by the California Department of Water and Power (DWP). It is situated on Site Class D soil as defined by ASCE 7-22. The ground motion was processed by applying baseline correction and low- and high-pass filters. The lowest usable frequency is 0.41 Hz, or the longest usable period is $T_u = 1/0.41 = 2.4$ s. Figure 6-10 shows acceleration, velocity, and displacement histories of Sylmar ground motion in maximum-horizontal (RotD100) directions. The peak ground acceleration, velocity, and displacement in RotD100 directions are $PGA = 0.964g$, $PGV = 52.1$ in./s, and $PGD = 22.3$ in. The number of cycles in velocity history is 1.4. With $ED = 15.4$ $(\text{m/s})^2$, $T_{cg} = 1.54$ s, and $PGV_n = 0.57$, Sylmar ground motion is extremely strong, medium to low frequency, and medium banded.

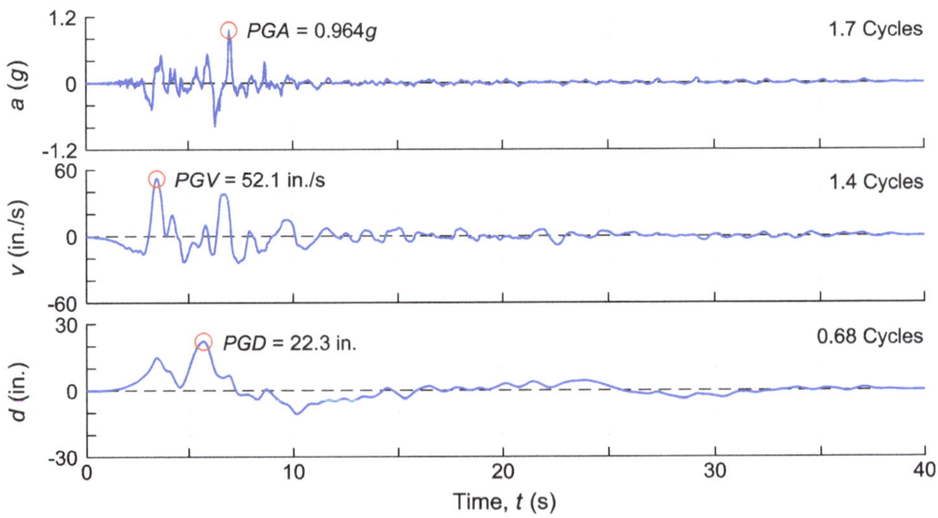

Figure 6-10. Acceleration, velocity, and displacement histories of Sylmar ground motion in maximum-horizontal (RotD100) directions.

6.5.1 Response of 3-Story Building to Sylmar Ground Motion

Per Table 5-1, the minimum *ED* required to collapse the 3-story steel SMF building is 23.8 $(m/s)^2$. The *ED* of Sylmar ground motion is only 15.4 $(m/s)^2$. It is centered close to the ultimate period of the building ($T_{cg} = 0.86\ T_{ult}$), but it is not narrowly focused ($PGV_n = 0.57$). Cyclic demand is moderate. Therefore, Sylmar ground motion cannot collapse the 3-story building, but it can damage the building. The response of the building is calculated from both static and dynamic methods of nonlinear analysis described in Chapter 4.

Figure 6-11 shows the demand-capacity equilibrium for the building subjected to Sylmar ground motion. Two demand curves are shown: (1) The magenta curve is the hypothetical demand curve without any hysteretic damping; it is the 2% damping ADRS; and (2) the black curve is the actual demand-curve for the building generated by the procedure described in Section 4.2. Per static analysis, the CG of the building deflects 18 in. The effective period of the building at equilibrium is $T_{eq} = 1.5$ s, and effective damping at equilibrium is $\zeta_{eq} = 17\%$. Using T_{eq} and ζ_{eq}, the equivalent number of cycles, read from cyclic-demand spectra (not shown), is $N_{eq} = 2$. Therefore, the CG of the building deflects 18 in. for two cycles. It was established in Chapter 3 that the CG of the building has the capacity to deflect 27 in. for two cycles. Therefore, the fatigue damage suffered by the building is $(18^2 \times 2)/(27^2 \times 2) = 0.45$ (45%) per static analysis. Table 6-7 provides a summary

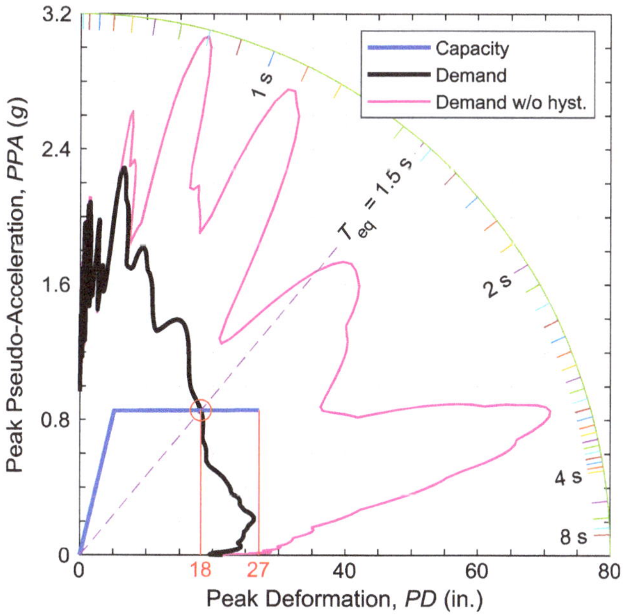

Figure 6-11. Demand-capacity equilibrium for 3-story building subjected to Sylmar ground motion shown in Figure 6-10.

Table 6-7. Response of 3-Story Building to Sylmar Ground Motion.

Nonlinear analysis	Roof deflection (in.)	Maximum drift ratio (radian)	Number of cycles	Fatigue damage (%)
Static	30.3	0.0701	2	45
Dynamic	22.5	0.0623	1.2	22

of results from static and dynamic analyses. The results from static analysis are more conservative in this case. If the building had no hysteretic damping (magenta curve in Figure 6-11), it would collapse because the capacity curve stops short of the magenta demand-curve.

6.5.2 Response of 20-Story Building to Sylmar Ground Motion

Per Table 5-1, the minimum *ED* required to collapse the 20-story steel EBF building is 34.1 $(m/s)^2$. The *ED* of Sylmar ground motion is only 15.4 $(m/s)^2$, and it is centered away from the ultimate period of the building ($T_{cg} = 0.33\ T_{ult}$). Cyclic demand is moderate. Therefore, Sylmar ground motion cannot collapse the 20-story building, but it can damage the building. The response of the building is calculated from both static and dynamic methods of nonlinear analysis described in Chapter 4.

Figure 6-12 shows the demand-capacity equilibrium for the building subjected to Sylmar ground motion. Two demand curves are shown: (1) The magenta curve is the

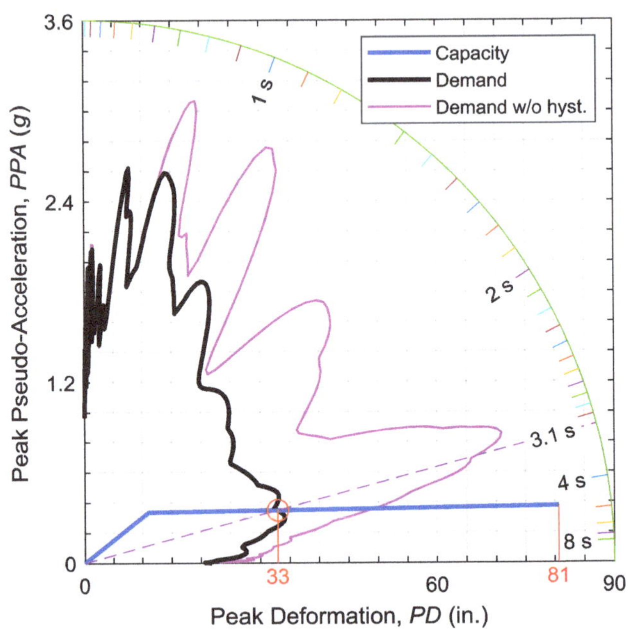

Figure 6-12. Demand-capacity equilibrium for 20-story building subjected to Sylmar ground motion shown in Figure 6-10.

Table 6-8. Response of 20-Story Building to Sylmar Ground Motion.

Nonlinear analysis	Roof deflection (in.)	Maximum drift ratio (radian)	Number of cycles	Fatigue damage (%)
Static	64.2	0.0223	2.5	42
Dynamic	31.6	0.0156	1.5	12

hypothetical demand curve without any hysteretic damping; it is the 2% damping ADRS; and (2) the black curve is the actual demand curve for the building generated by the procedure described in Section 4.2. Per static analysis, the CG of the building deflects 33 in. The effective period of the building at equilibrium is $T_{eq} = 3.1$ s, and effective damping at equilibrium is $\zeta_{eq} = 14\%$. Using T_{eq} and ζ_{eq}, the equivalent number of cycles, read from cyclic-demand spectra (not shown), is $N_{eq} = 2.5$. Therefore, the CG of the building deflects 33 in. for 2.5 cycles. It was established in Chapter 3 that the CG of the building has the capacity to deflect 81 in. for one cycle. Therefore, the fatigue damage suffered by the building is $(33^2 \times 2.5)/(81^2 \times 1) = 0.42$ (42%). Notice in Figure 6-12 that the high-frequency energy, associated with high *PGA*, is not absorbed by the building. Table 6-8 provides a summary of results from static and dynamic analyses. The results from static analysis are more conservative in this case. If the building had no hysteretic damping (magenta curve in Figure 6-12), it would deflect 49 in. for 3.5 cycles and suffer 66% fatigue damage.

6.6 Chi-Chi, Taiwan, CHY080 Ground Motion

This ground motion was recorded at CHY080 station, at 0.11 km from vertical projection of the fault, during the 1999 magnitude **M** 7.6 Chi-Chi, Taiwan Earthquake. The CHY080 station is owned and maintained by the Central Weather Bureau of Taiwan. Per PEER (2022), the CHY080 station is situated on Site Class C soil as defined by ASCE 7-22 (ASCE 2022). The data were processed by applying baseline correction and low- and high-pass filters (PEER 2022). The lowest usable frequency is 0.125 Hz, or the longest usable period is $T_u = 1/0.125 = 8$ s. Figure 6-13 shows acceleration, velocity, and displacement histories of CHY080 ground motion in maximum-horizontal (RotD100) directions. The peak ground acceleration, velocity, and displacement in RotD100 directions are $PGA = 1.1g$, $PGV = 42.8$ in./s, and $PGD = 14.9$ in. The number of cycles in velocity history is 2.1. With $ED = 15.9$ (m/s)^2, $T_{cg} = 1.01$ s, and $PGV_n = 0.54$, CHY080 ground motion is extremely strong, medium frequency, and medium banded.

6.6.1 Response of 3-Story Building to CHY080 Ground Motion

Per Table 5-1, the minimum *ED* required to collapse the 3-story steel SMF building is 23.8 (m/s)^2. The *ED* of CHY080 ground motion is 15.9 (m/s)^2, and it is centered away from the ultimate period of the building ($T_{cg} = 0.56\ T_{ult}$). Cyclic demand is

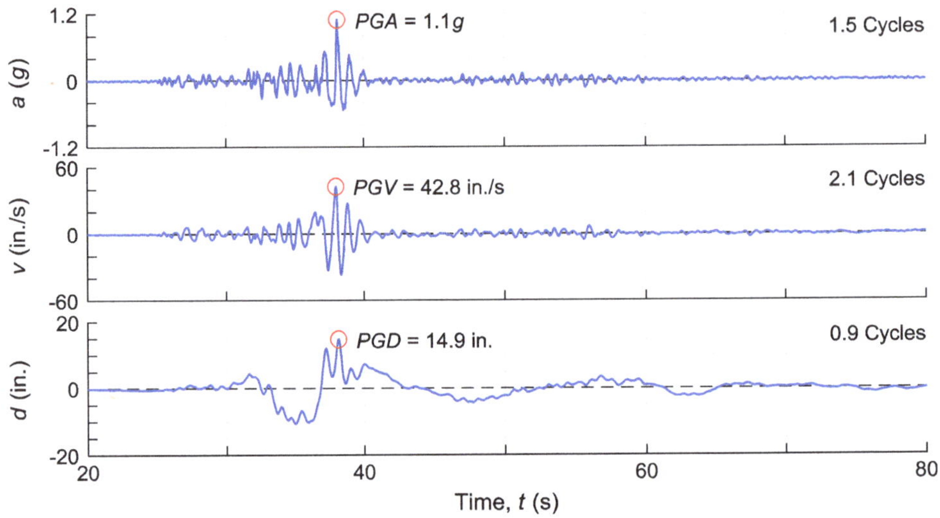

Figure 6-13. Acceleration, velocity, and displacement histories of CHY080 ground motion in maximum-horizontal (RotD100) directions.

moderate. Therefore, CHY080 ground motion cannot collapse the 3-story building, but it can damage the building. The response of the building to CHY080 ground motion is calculated from both static and dynamic methods of nonlinear analysis described in Chapter 4.

Figure 6-14 shows the demand-capacity equilibrium for the building subjected to CHY080 ground motion. Two demand curves are shown: (1) The magenta curve is the hypothetical demand curve without any hysteretic damping; it is the 2% damping ADRS; and (2) the black curve is the actual demand curve for the building generated by the procedure described in Section 4.2. Per static analysis, the CG of the building deflects 18 in. The effective period of the building at equilibrium is $T_{eq} = 1.5$ s, and effective damping at equilibrium is $\zeta_{eq} = 17\%$. Using T_{eq} and ζ_{eq}, the equivalent number of cycles, read from cyclic-demand spectra (not shown), is $N_{eq} = 1.5$. Therefore, the CG of the building deflects 18 in. for 1.5 cycles. It was established in Chapter 3 that the CG of the building has the capacity to deflect 27 in. for two cycles. Therefore, the fatigue damage suffered by the building is $(18^2 \times 1.5)/(27^2 \times 2) = 0.32$ (32%) per static analysis. Table 6-9 provides a summary of results from static and dynamic analyses. The results from static analysis are more conservative in this case. If the building had no hysteretic damping (magenta curve in Figure 6-14), it would deflect 26 in. for 2.5 cycles and suffer 100% fatigue damage, resulting in collapse.

6.6.2 Response of 20-Story Building to CHY080 Ground Motion

Per Table 5-1, the minimum *ED* required to collapse the 20-story steel EBF building is 34.1 $(m/s)^2$. The *ED* of CHY080 ground motion is only 15.9 $(m/s)^2$, and it is centered

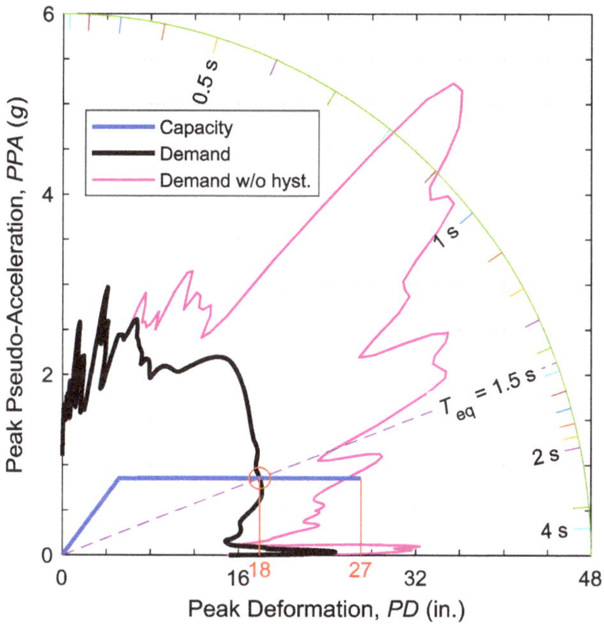

Figure 6-14. Demand-capacity equilibrium for 3-story building subjected to CHY080 ground motion shown in Figure 6-13.

Table 6-9. Response of 3-Story Building to CHY080 Ground Motion.

Nonlinear analysis	Roof deflection (in.)	Maximum drift ratio (radian)	Number of cycles	Fatigue damage (%)
Static	29.8	0.0690	1.5	32
Dynamic	21.8	0.0627	1.7	30

away from the ultimate period of the building ($T_{cg}=0.21\ T_{ult}$). Cyclic demand is moderate. Therefore, CHY080 ground motion cannot collapse the 20-story building, but it can damage the building. The response of the building is calculated from both static and dynamic methods of nonlinear analysis described in Chapter 4.

Figure 6-15 shows the demand-capacity equilibrium for the building subjected to CHY080 ground motion. Two demand curves are shown: (1) The magenta curve is the hypothetical demand curve without any hysteretic damping; it is the 2% damping ADRS; and (2) the black curve is the actual demand curve for the building generated by the procedure described in Section 4.2. Per static analysis, the CG of the building deflects 19 in. The effective period of the building at equilibrium is $T_{eq}=2.4$ s, and effective damping at equilibrium is $\zeta_{eq}=6.3\%$. Using T_{eq} and ζ_{eq}, the equivalent

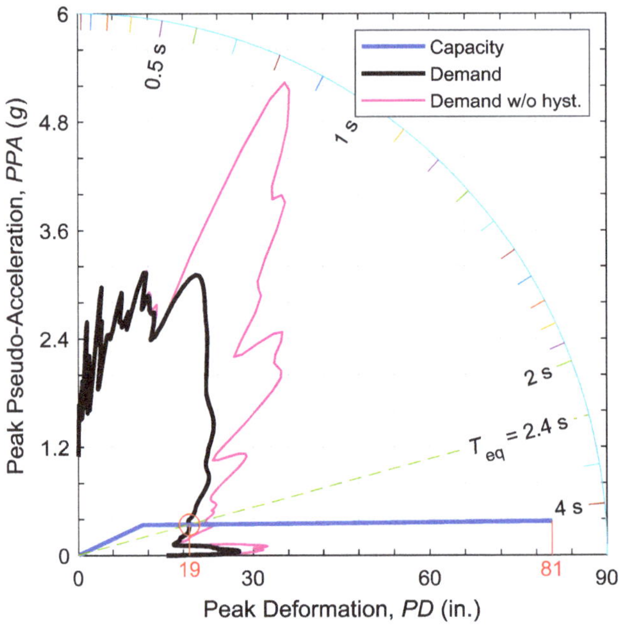

Figure 6-15. Demand-capacity equilibrium for 20-story building subjected to CHY080 ground motion shown in Figure 6-13.

number of cycles, read from cyclic-demand spectra (not shown), is $N_{eq} = 1.1$. Therefore, the CG of the building deflects 19 in. for 1.1 cycles. It was established in Chapter 3 that the CG of the building has the capacity to deflect 81 in. for one cycle. Therefore, the fatigue damage suffered by the building is $(19^2 \times 1.1)/(81^2 \times 1) = 0.06$ (6%). Notice in Figure 6-15 that the high-frequency energy, associated with high *PGA*, is not absorbed by the building. Table 6-10 provides a summary of results from static and dynamic analyses. The results from dynamic analysis are more conservative in this case. Per dynamic analysis, the building suffers 16% fatigue damage. If the building had no hysteretic damping (magenta curve in Figure 6-15), it would deflect 23 in. for 4.7 cycles and suffer 39% fatigue damage.

Table 6-10. Response of 20-Story Building to CHY080 Ground Motion.

Nonlinear analysis	Roof deflection (in.)	Maximum drift ratio (radian)	Number of cycles	Fatigue damage (%)
Static	37.2	0.0129	1.1	6
Dynamic	24.7	0.0156	1.9	16

6.7 Kobe, Takatori Ground Motion

This ground motion was recorded at the Takatori station, at 1.46 km from vertical projection of the fault, during the 1995 magnitude **M** 6.8 Kobe, Japan Earthquake. The Takatori station is owned and maintained by the Railway Technical Research Institute of Japan. Per PEER (2022), the Takatori station is situated on Site Class D soil as defined by ASCE 7-22. The data used in this study were processed by applying baseline correction and low- and high-pass filters. The lowest usable frequency is 0.356 Hz, or the longest usable period is $T_u = 1/0.356 = 2.8$ s. Figure 6-16 shows acceleration, velocity, and displacement histories of Takatori ground motion in maximum-horizontal (RotD100) directions. The peak ground acceleration, velocity, and displacement in RotD100 directions are $PGA = 0.715g$, $PGV = 63.1$ in./s, and $PGD = 17.1$ in. The number of cycles in velocity history is 1.5. With $ED = 20$ $(\text{m/s})^2$, $T_{cg} = 1.44$ s, and $PGV_n = 0.92$, Takatori ground motion is extremely strong, medium frequency, and narrow banded.

6.7.1 Response of 3-Story Building to Takatori Ground Motion

Per Table 5-1, the minimum *ED* required to collapse the 3-story steel SMF building is 23.8 $(\text{m/s})^2$. The *ED* of Takatori ground motion is 20 $(\text{m/s})^2$. It is centered close to the ultimate period of the building ($T_{cg} = 0.8\ T_{ult}$), and it is narrowly focused ($PGV_n = 0.92$). Therefore, Takatori ground motion could collapse the 3-story building if it has sufficient cyclic demand. The response of the building is calculated from both static and dynamic methods of nonlinear analysis described in Chapter 4.

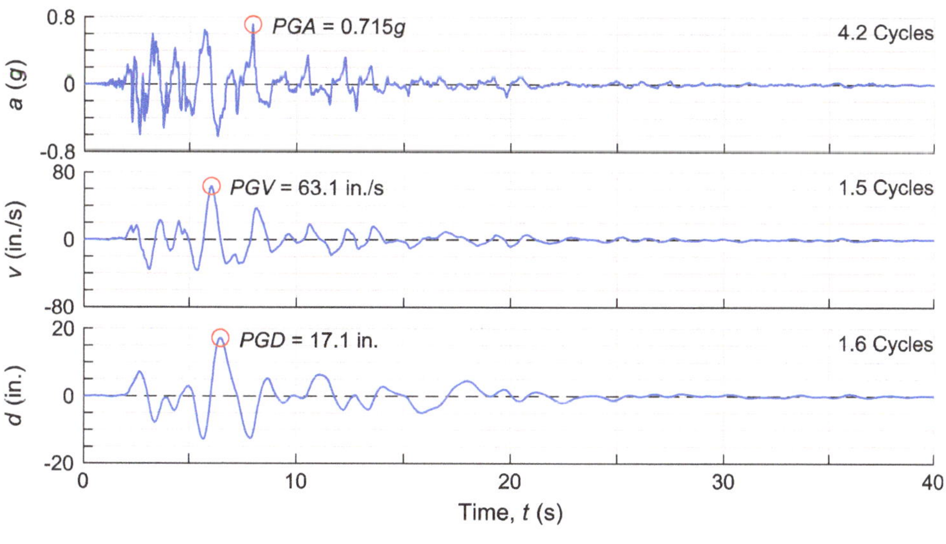

Figure 6-16. Acceleration, velocity, and displacement histories of Takatori ground motion in maximum-horizontal (RotD100) directions.

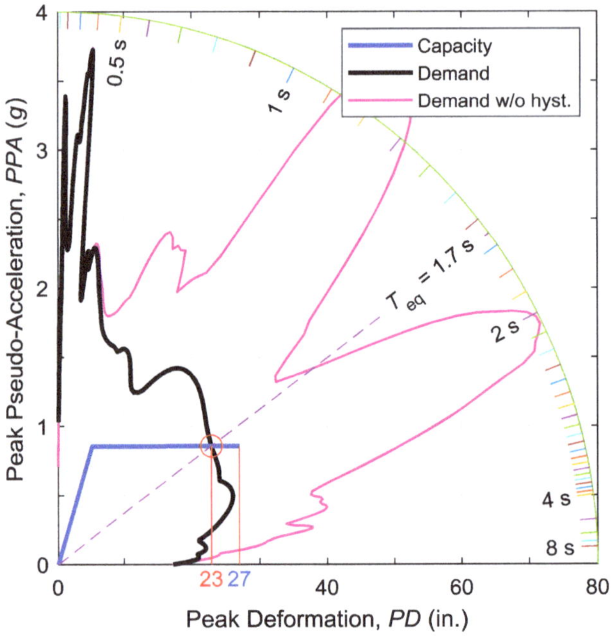

Figure 6-17. Demand-capacity equilibrium for 3-story building subjected to Takatori ground motion shown in Figure 6-16.

Figure 6-17 shows the demand-capacity equilibrium for the building subjected to Takatori ground motion. Two demand curves are shown: (1) The magenta curve is the hypothetical demand curve without any hysteretic damping; it is the 2% damping ADRS; and (2) the black curve is the actual demand curve for the building generated by the procedure described in Section 4.2. Per static analysis, the CG of the building deflects 23 in. The effective period of the building at equilibrium is $T_{eq} = 1.7$ s, and effective damping at equilibrium is $\zeta_{eq} = 20\%$. Using T_{eq} and ζ_{eq}, the equivalent number of cycles, read from cyclic-demand spectra (not shown), is $N_{eq} = 1.9$. Therefore, the CG of the building deflects 23 in. for 1.9 cycles. It was established in Chapter 3 that the CG of the building has the capacity to deflect 27 in. for two cycles. Therefore, the fatigue damage suffered by the building is $(23^2 \times 1.9)/(27^2 \times 2) = 0.68$ (68%) per static analysis. Table 6-11 provides a summary of results from static and dynamic analyses. The results from static analysis are more

Table 6-11. Response of 3-Story Building to Takatori Ground Motion.

Nonlinear analysis	Roof deflection (in.)	Maximum drift ratio (radian)	Number of cycles	Fatigue damage (%)
Static	38.2	0.0884	1.9	68
Dynamic	21.5	0.0602	1.5	25

conservative in this case. If the building had no hysteretic damping (magenta curve in Figure 6-17), it would collapse because the capacity curve stops short of the magenta demand curve.

6.7.2 Response of 20-Story Building to Takatori Ground Motion

Per Table 5-1, the minimum *ED* required to collapse the 20-story steel EBF building is 34.1 $(m/s)^2$. The *ED* of Takatori ground motion is only 20 $(m/s)^2$ and it is centered away from the ultimate period of the building ($T_{cg} = 0.31\ T_{ult}$). Cyclic demand is moderate. Therefore, Takatori ground motion cannot collapse the 20-story building, but it can damage the building. The response of the building to Takatori ground motion is calculated from both static and dynamic methods of nonlinear analysis described in Chapter 4.

Figure 6-18 shows the demand-capacity equilibrium for the building subjected to Takatori ground motion. Two demand curves are shown: (1) The magenta curve is the hypothetical demand curve without any hysteretic damping; it is the 2% damping ADRS; and (2) the black curve is the actual demand curve for the building generated by the procedure described in Section 4.2. Per static analysis, the CG of the building deflects 28 in. The effective period of the building at equilibrium is $T_{eq} = 2.9$ s, and

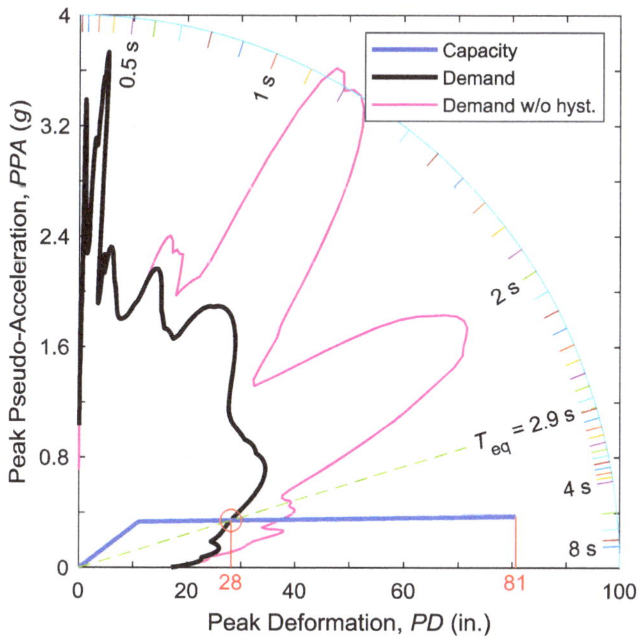

Figure 6-18. Demand-capacity equilibrium for 20-story building subjected to Takatori ground motion shown in Figure 6-16.

Table 6-12. Response of 20-Story Building to Takatori Ground Motion.

Nonlinear analysis	Roof deflection (in.)	Maximum drift ratio (radian)	Number of cycles	Fatigue damage (%)
Static	54.9	0.0191	2.2	26
Dynamic	44	0.0257	1.3	30

effective damping at equilibrium is $\zeta_{eq} = 12\%$. Using T_{eq} and ζ_{eq}, the equivalent number of cycles, read from cyclic-demand spectra (not shown), is $N_{eq} = 2.2$. Therefore, the CG of the building deflects 28 in. for 2.2 cycles. It was established in Chapter 3 that the CG of the building has the capacity to deflect 81 in. for one cycle. Therefore, the fatigue damage suffered by the building is $(28^2 \times 2.2)/(81^2 \times 1) = 0.26$ (26%). Notice in Figure 6-18 that the high-frequency energy (associated with large accelerations) is not absorbed by the building. Since $T_{eq} > T_u$, the results of analysis are less reliable. Table 6-12 provides a summary of results from static and dynamic analyses. The results from dynamic analysis are more conservative in this case. Per dynamic analysis, the building suffers 30% fatigue damage. If the building had no hysteretic damping (magenta curve in Figure 6-18), it would deflect 37 in. for 4.6 cycles and suffer 100% fatigue damage, resulting in collapse.

6.8 Chi-Chi, Taiwan, TCU084 Ground Motion

This ground motion was recorded at the TCU084 station, at zero distance from vertical projection of the fault, during the 1999 magnitude **M** 7.6 Chi-Chi, Taiwan Earthquake. The TCU084 station is owned and maintained by the Central Weather Bureau of Taiwan. Per PEER (2022), the station is situated on Site Class C soil as defined by ASCE 7-22. The data used in this study were processed by applying baseline correction and low- and high-pass filters (PEER 2022). The lowest usable frequency is 0.25 Hz, or the longest usable period is $T_u = 1/0.25 = 4$ s. Figure 6-19 shows acceleration, velocity, and displacement histories of TCU084 ground motion in maximum-horizontal (RotD100) directions. The peak ground acceleration, velocity, and displacement in RotD100 directions are $PGA = 1.02g$, $PGV = 45.7$ in./s, and $PGD = 15.1$ in. The number of cycles in velocity history is 3.8. With $ED = 20$ (m/s)^2, $T_{cg} = 1.28$ s, and $PGV_n = 0.59$, TCU084 ground motion is extremely strong, medium frequency, and medium banded.

6.8.1 Response of 3-Story Building to TCU084 Ground Motion

Per Table 5-1, the minimum *ED* required to collapse the 3-story steel SMF building is 23.8 (m/s)^2. The *ED* of TCU084 ground motion is 20 (m/s)^2. It is centered close to the ultimate period of the building ($T_{cg} = 0.71\ T_{ult}$), but it is not narrowly focused ($PGV_n = 0.59$). Cyclic demand is high for this ground motion. TCU084 ground

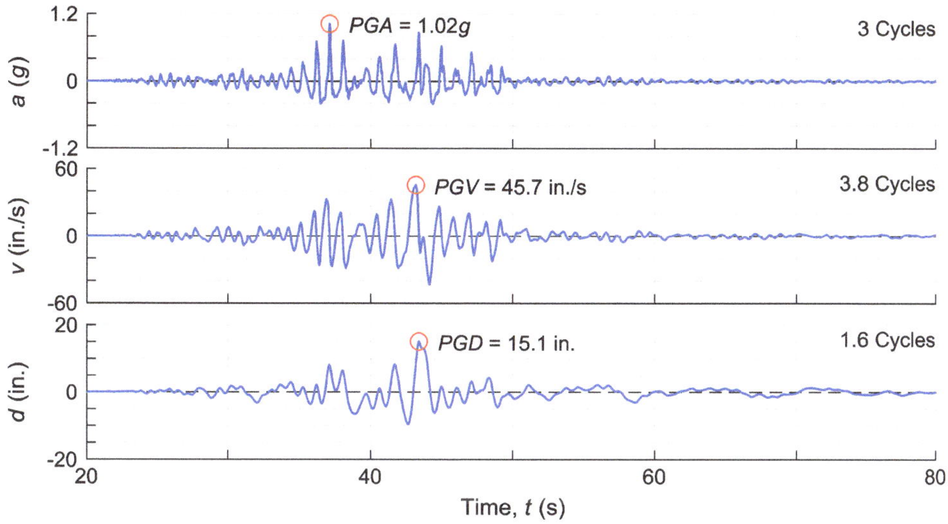

Figure 6-19. Acceleration, velocity, and displacement histories of TCU084 ground motion in maximum-horizontal (RotD100) directions.

motion is not likely to collapse the 3-story building, but it can cause significant damage. The response of the building to TCU084 ground motion is calculated from both static and dynamic methods of nonlinear analysis described in Chapter 4.

Figure 6-20 shows the demand-capacity equilibrium for the building subjected to TCU084 ground motion. Two demand curves are shown: (1) The magenta curve is the hypothetical demand curve without any hysteretic damping; it is the 2% damping ADRS; and (2) the black curve is the actual demand curve for the building generated by the procedure described in Section 4.2. Per static analysis, the CG of the building deflects 18 in. The effective period of the building at equilibrium is $T_{eq} = 1.5$ s, and effective damping at equilibrium is $\zeta_{eq} = 17\%$. Using T_{eq} and ζ_{eq}, the equivalent number of cycles, read from cyclic-demand spectra (not shown), is $N_{eq} = 3.5$. Therefore, the CG of the building deflects 18 in. for 3.5 cycles. It was established in Chapter 3 that the CG of the building has the capacity to deflect 27 in. for two cycles. Therefore, the fatigue damage suffered by the building is $(18^2 \times 3.5)/(27^2 \times 2) = 0.81$ (81%) per static analysis. Very high damage caused by this ground motion is because of its high cyclic demand. Table 6-13 provides a summary of results from static and dynamic analyses. The results from static analysis are more conservative in this case. If the building had no hysteretic damping (magenta curve in Figure 6-20), it would collapse because the capacity curve stops short of the magenta demand curve.

6.8.2 Response of 20-Story Building to TCU084 Ground Motion

Per Table 5-1, the minimum *ED* required to collapse the 20-story steel EBF building is 34.1 $(m/s)^2$. The *ED* of TCU084 ground motion is only 20 $(m/s)^2$, and it is centered

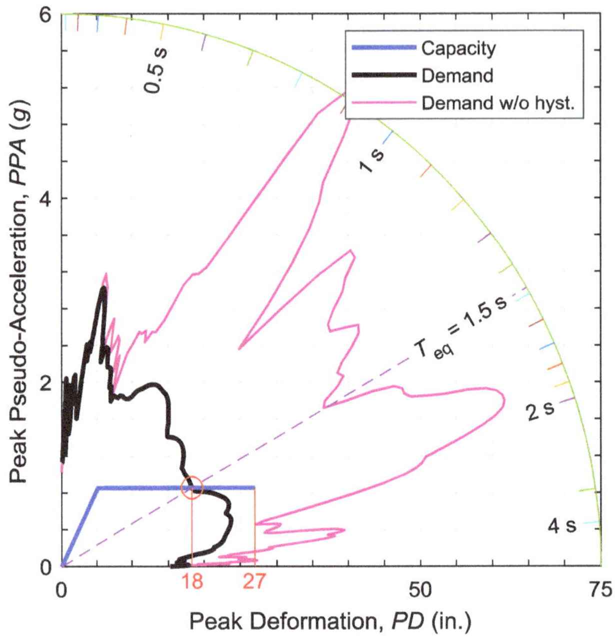

Figure 6-20. Demand-capacity equilibrium for 3-story building subjected to TCU084 ground motion shown in Figure 6-19.

Table 6-13. Response of 3-Story Building to TCU084 Ground Motion.

Nonlinear analysis	Roof deflection (in.)	Maximum drift ratio (radian)	Number of cycles	Fatigue damage (%)
Static	30.5	0.0707	3.5	81
Dynamic	26.8	0.0700	2.4	55

away from the ultimate period of the building ($T_{cg} = 0.27\ T_{ult}$). Despite high cyclic demand, TCU084 ground motion is not expected to collapse the 20-story building. The response of the building is calculated from both static and dynamic methods of nonlinear analysis described in Chapter 4.

Figure 6-21 shows the demand-capacity equilibrium for the building subjected to TCU084 ground motion. Two demand curves are shown: (1) The magenta curve is the hypothetical demand curve without any hysteretic damping; it is the 2% damping ADRS; and (2) the black curve is the actual demand curve for the building generated by the procedure described in Section 4.2. Per static analysis, the CG of the building deflects 26 in. The effective period of the building at equilibrium is $T_{eq} = 2.8$ s, and effective damping at equilibrium is $\zeta_{eq} = 11\%$. Using T_{eq} and ζ_{eq}, the equivalent

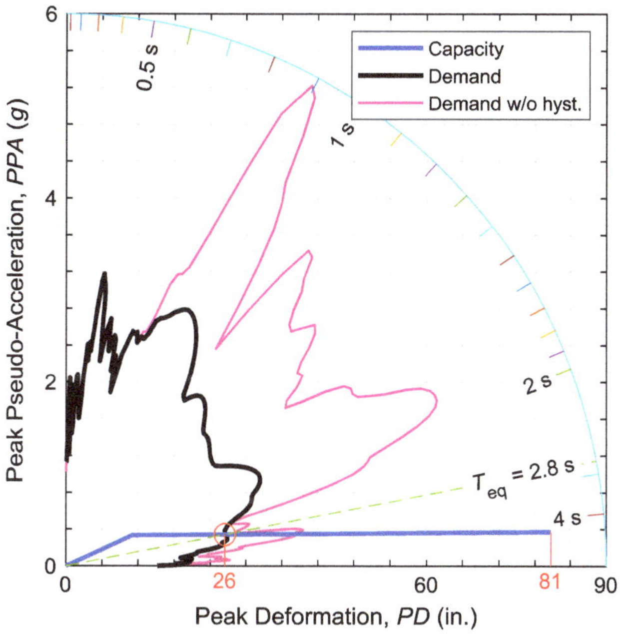

Figure 6-21. Demand-capacity equilibrium for 20-story building subjected to TCU084 ground motion shown in Figure 6-19.

number of cycles, read from cyclic-demand spectra (not shown), is $N_{eq} = 1.8$. Therefore, the CG of the building deflects 26 in. for 1.8 cycles. It was established in Chapter 3 that the CG of the building has the capacity to deflect 81 in. for one cycle. Therefore, the fatigue damage suffered by the building is $(26^2 \times 1.8)/(81^2 \times 1) = 0.19$ (19%). Notice in Figure 6-21 that the high-frequency energy, associated with high *PGA*, is not absorbed by the building. Table 6-14 provides a summary of results from static and dynamic analyses. The results from dynamic analysis are more conservative in this case. Per dynamic analysis, the building suffers 27% fatigue damage. If the building had no hysteretic damping (magenta curve in Figure 6-21), it would deflect 37 in. for 5.8 cycles and suffer 100% fatigue damage, resulting in collapse.

Table 6-14. Response of 20-Story Building to TCU084 Ground Motion.

Nonlinear analysis	Roof deflection (in.)	Maximum drift ratio (radian)	Number of cycles	Fatigue damage (%)
Static	51.6	0.0179	1.8	19
Dynamic	31.1	0.0171	2.8	27

6.9 Nurdagi, Türkiye, TK3123 Ground Motion

This ground motion was recorded at TK3123 station, during the 2023 magnitude **M** 7.8 Nurdagi, Türkiye Earthquake. The TK3123 station is owned and maintained by the Disaster and Emergency Management Presidency of Türkiye (AFAD 2023). Per AFAD (2023), the station is situated on Site Class C soil as defined by ASCE 7-22. The data used in this study were processed by AFAD (2023) by applying baseline correction and low- and high-pass filters. The lowest usable frequency is 0.025 Hz, or the longest usable period is T_u = 1/0.025 = 40 s. Figure 6-22 shows acceleration, velocity, and displacement histories of TK3123 ground motion in maximum-horizontal (RotD100) directions. The peak ground acceleration, velocity, and displacement in RotD100 directions are PGA = 0.665g, PGV = 73.5 in./s, and PGD = 36.5 in. The number of cycles in velocity history is 1.3. With ED = 20.1 (m/s)^2, T_{cg} = 1.84 s, and PGV_n = 0.76, TK3123 ground motion is extremely strong, medium to low frequency, and narrow banded. TK3123 ground motion is the strongest ground motion recorded during the 2023 Nurdagi, Türkiye Earthquake.

6.9.1 Response of 3-Story Building to TK3123 Ground Motion

Per Table 5-1, the minimum ED required to collapse the 3-story steel SMF building is 23.8 (m/s)^2. The ED of TK3123 ground motion is 20.1 (m/s)^2. It is centered very close to the ultimate period of the building ($T_{cg} \approx T_{ult}$), and it is also narrowly focused (PGV_n = 0.76). However, cyclic demand is low. TK3123 ground motion is not expected to collapse the 3-story building, but it can cause significant damage.

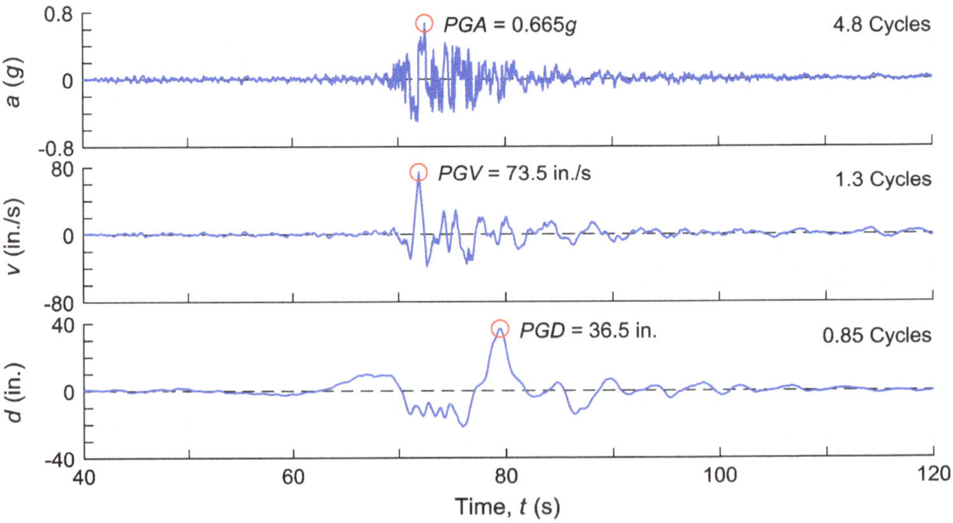

Figure 6-22. Acceleration, velocity, and displacement histories of TK3123 ground motion in maximum-horizontal (RotD100) directions.

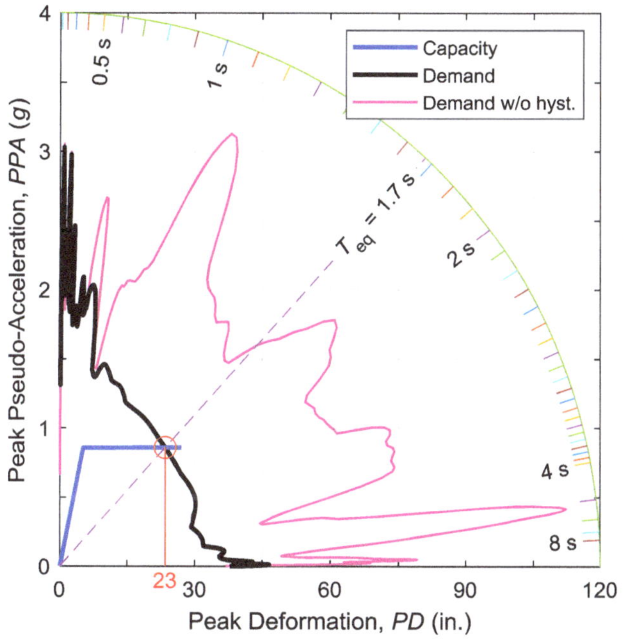

Figure 6-23. Demand-capacity equilibrium for 3-story building subjected to TK3123 ground motion shown in Figure 6-22.

The response of the building is calculated from both static and dynamic methods of nonlinear analysis described in Chapter 4.

Figure 6-23 shows the demand-capacity equilibrium for the building subjected to TK3123 ground motion. Two demand curves are shown: (1) The magenta curve is the hypothetical demand curve without any hysteretic damping; it is the 2% damping ADRS; and (2) the black curve is the actual demand-curve for the building generated by the procedure described in Section 4.2. Per static analysis, the CG of the building deflects 23 in. The effective period of the building at equilibrium is $T_{eq} = 1.7$ s, and effective damping at equilibrium is $\zeta_{eq} = 21\%$. Using T_{eq} and ζ_{eq}, the equivalent number of cycles, read from cyclic-demand spectra (not shown), is $N_{eq} = 1.5$. Therefore, the CG of the building deflects 23 in. for 1.5 cycles. It was established in Chapter 3 that the CG of the building has the capacity to deflect 27 in. for two cycles. Therefore, the fatigue damage suffered by the building is $(23^2 \times 1.5)/(27^2 \times 2) = 0.58$ (58%) per static analysis. Lower cyclic demand is the reason for less than expected damage for this ground motion. Table 6-15 provides a summary of results from static and dynamic analyses. The results from static analysis are more conservative in this case. If the building had no hysteretic damping (magenta curve in Figure 6-23), it would collapse because the capacity curve stops short of the magenta demand curve.

Table 6-15. Response of 3-Story Building to TK3123 Ground Motion.

Nonlinear analysis	Roof deflection (in.)	Maximum drift ratio (radian)	Number of cycles	Fatigue damage (%)
Static	39.1	0.0906	1.5	58
Dynamic	15.5	0.0471	2.3	24

6.9.2 Response of 20-Story Building to TK3123 Ground Motion

Per Table 5-1, the minimum *ED* required to collapse the 20-story steel EBF building is 34.1 $(m/s)^2$. The *ED* of TK3123 ground motion is 20.1 $(m/s)^2$, and it is centered away from the ultimate period of the building ($T_{cg} = 0.39\ T_{ult}$). Cyclic demand is also low. Therefore, TK3123 ground motion cannot collapse the 20-story building, but it can damage the building. The response of the building is calculated from both static and dynamic methods of nonlinear analysis described in Chapter 4.

Figure 6-24 shows the demand-capacity equilibrium for the building subjected to TK3123 ground motion. Two demand curves are shown: (1) The magenta curve is the hypothetical demand curve without any hysteretic damping; it is the 2% damping ADRS; and (2) the black curve is the actual demand curve for the building generated by the procedure described in Section 4.2. Per static analysis, the CG of the building deflects 36 in. The effective period of the building at equilibrium is $T_{eq} = 3.2$ s, and effective damping at equilibrium is $\zeta_{eq} = 15\%$. Using T_{eq} and ζ_{eq}, the equivalent

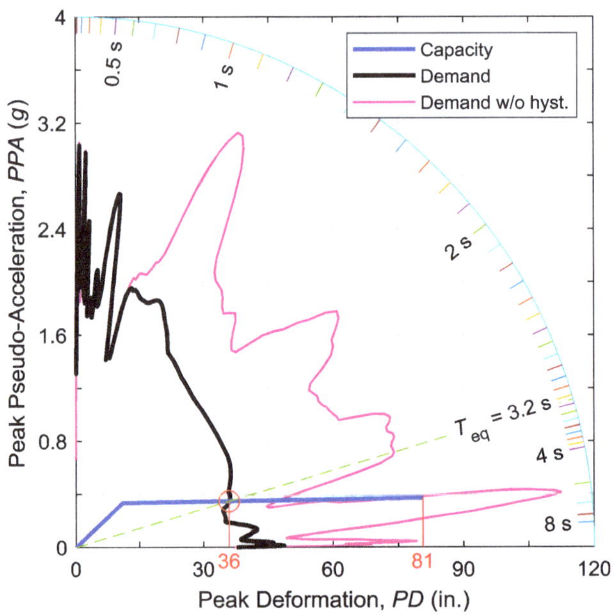

Figure 6-24. Demand-capacity equilibrium for 20-story building subjected to TK3123 ground motion shown in Figure 6-22.

Table 6-16. Response of 20-Story Building to TK3123 Ground Motion.

Nonlinear analysis	Roof deflection (in.)	Maximum drift ratio (radian)	Number of cycles	Fatigue damage (%)
Static	70.1	0.0243	2.3	46
Dynamic	39.5	0.0217	1.7	27

number of cycles, read from cyclic-demand spectra (not shown), is $N_{eq} = 2.3$. Therefore, the CG of the building deflects 36 in. for 2.3 cycles. It was established in Chapter 3 that the CG of the building has the capacity to deflect 81 in. for one cycle. Therefore, the fatigue damage suffered by the building is $(36^2 \times 2.3)/(81^2 \times 1) = 0.46$ (46%). Notice in Figure 6-24 that the high-frequency energy, associated with high *PGA*, is not absorbed by the building. Table 6-16 provides a summary of results from static and dynamic analyses. The results from static analysis are more conservative in this case. If the building had no hysteretic damping (magenta curve in Figure 6-24), it would deflect 47 in. for three cycles and suffer 100% fatigue damage, resulting in collapse.

6.10 Chi-Chi, Taiwan, TCU052 Ground Motion

This ground motion was recorded at TCU052 station, at zero distance from vertical projection of the fault, during the 1999 magnitude **M** 7.6 Chi-Chi, Taiwan Earthquake. The TCU052 station is owned and maintained by the Central Weather Bureau of Taiwan. Per PEER (2022), the station is situated on Site Class C soil as defined by ASCE 7-22. The data used in this study were processed by applying baseline correction and low- and high-pass filters (PEER 2022). The lowest usable frequency is 0.05 Hz, or the longest usable period is $T_u = 1/0.05 = 20$ s. Figure 6-25 shows acceleration, velocity, and displacement histories of TCU052 ground motion in maximum-horizontal (RotD100) directions. The peak ground acceleration, velocity, and displacement in RotD100 directions are $PGA = 0.458g$, $PGV = 70.6$ in./s, and $PGD = 101$ in. The number of cycles in velocity history is 1.2. With $ED = 21.4$ $(\text{m/s})^2$, $T_{cg} = 3.95$ s, and $PGV_n = 0.53$, TCU052 ground motion is extremely strong, low frequency, and medium banded.

6.10.1 Response of 3-Story Building to TCU052 Ground Motion

Per Table 5-1, the minimum *ED* required to collapse the 3-story steel SMF building is 23.8 $(\text{m/s})^2$. The *ED* of TCU052 ground motion is 21.4 $(\text{m/s})^2$, and it is not centered close to the ultimate period of the building ($T_{cg} = 2.2\ T_{ult}$). Cyclic demand is also low. Therefore, TCU052 ground motion cannot collapse the 3-story building, but it can damage the building. The response of the building is calculated from both static and dynamic methods of nonlinear analysis described in Chapter 4.

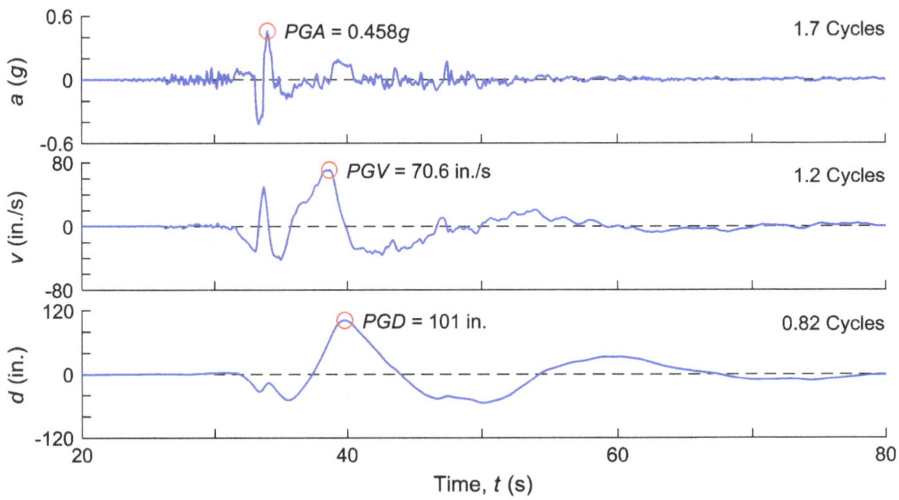

Figure 6-25. Acceleration, velocity, and displacement histories of TCU052 ground motion in maximum-horizontal (RotD100) directions.

Figure 6-26 shows the demand-capacity equilibrium for the building subjected to TCU052 ground motion. Two demand curves are shown: (1) The magenta curve is the hypothetical demand curve without any hysteretic damping; it is the 2% damping ADRS; and (2) the black curve is the actual demand curve for the building generated

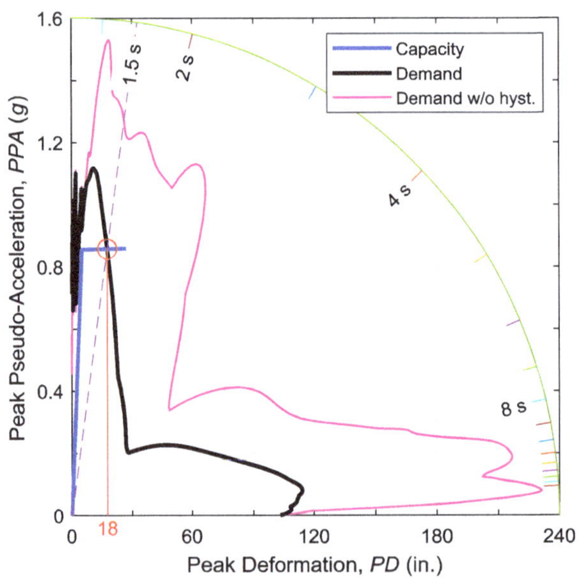

Figure 6-26. Demand-capacity equilibrium for 3-story building subjected to TCU052 ground motion shown in Figure 6-25.

Table 6-17. Response of 3-Story Building to TCU052 Ground Motion.

Nonlinear analysis	Roof deflection (in.)	Maximum drift ratio (radian)	Number of cycles	Fatigue damage (%)
Static	29.4	0.0682	1.2	25
Dynamic	12.9	0.0403	1.2	9

by the procedure described in Section 4.2. Per static analysis, the CG of the building deflects 18 in. The effective period of the building at equilibrium is $T_{eq} = 1.5$ s, and effective damping at equilibrium is $\zeta_{eq} = 17\%$. Using T_{eq} and ζ_{eq}, the equivalent number of cycles, read from cyclic-demand spectra (not shown), is $N_{eq} = 1.2$. Therefore, the CG of the building deflects 18 in. for 1.2 cycles. It was established in Chapter 3 that the CG of the building has the capacity to deflect 27 in. for two cycles. Therefore, the fatigue damage suffered by the building is $(18^2 \times 1.2)/(27^2 \times 2) = 0.25$ (25%) per static analysis. Notice in Figure 6-26 that the low-frequency energy, associated with high *PGD*, is not absorbed by the building. Table 6-17 provides a summary of results from static and dynamic analyses. The results from static analysis are more conservative in this case. If the building had no hysteretic damping (magenta curve in Figure 6-26), it would collapse because the capacity curve stops short of the magenta demand curve.

6.10.2 Response of 20-Story Building to TCU052 Ground Motion

Per Table 5-1, the minimum *ED* required to collapse the 20-story steel EBF building is 34.1 $(m/s)^2$. The *ED* of TCU052 ground motion is only 21.4 $(m/s)^2$. It is centered close to the ultimate period of the building ($T_{cg} = 0.84\ T_{ult}$), but it is not narrowly focused ($PGV_n = 0.53$). Cyclic demand is also low. Therefore, TCU052 ground motion cannot collapse the 20-story building, but it can damage the building. The response of the building is calculated from both static and dynamic methods of nonlinear analysis described in Chapter 4.

Figure 6-27 shows the demand-capacity equilibrium for the building subjected to TCU052 ground motion. Two demand curves are shown: (1) The magenta curve is the hypothetical demand-curve without any hysteretic damping; it is the 2% damping ADRS; and (2) the black curve is the actual demand-curve for the building generated by the procedure described in Section 4.2. Per static analysis, the CG of the building deflects 34 in. The effective period of the building at equilibrium is $T_{eq} = 3.2$ s, and effective damping at equilibrium is $\zeta_{eq} = 15\%$. Using T_{eq} and ζ_{eq}, the equivalent number of cycles, read from cyclic-demand spectra (not shown), is $N_{eq} = 1.5$. Therefore, the CG of the building deflects 34 in. for 1.5 cycles. It was established in Chapter 3 that the CG of the building has the capacity to deflect 81 in. for one cycle. Therefore, the fatigue damage suffered by the building is

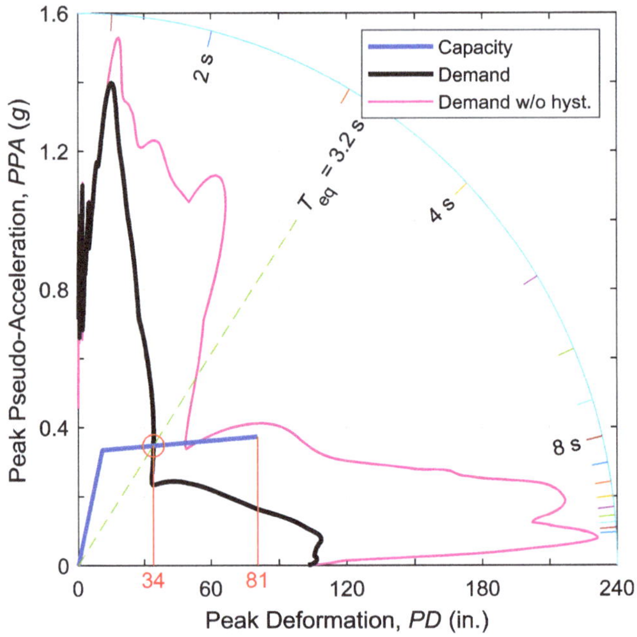

Figure 6-27. Demand-capacity equilibrium for 20-story building subjected to TCU052 ground motion shown in Figure 6-25.

Table 6-18. Response of 20-Story Building to TCU052 Ground Motion.

Nonlinear analysis	Roof deflection (in.)	Maximum drift ratio (radian)	Number of cycles	Fatigue damage (%)
Static	66.4	0.0231	1.5	27
Dynamic	35.5	0.0173	1.9	19

$(34^2 \times 1.5)/(81^2 \times 1) = 0.27$ (27%). Notice in Figure 6-27 that high-frequency energy, associated with high *PGA*, and low-frequency energy, associated with high *PGD*, are not absorbed by the building. Table 6-18 provides a summary of results from static and dynamic analyses. The results from static analysis are more conservative in this case. If the building had no hysteretic damping (magenta curve in Figure 6-27), it would deflect 48 in. for 2.2 cycles and suffer 79% fatigue damage.

6.11 Niigata, Japan, NIG019 Ground Motion

This ground motion was recorded at NIG019 station, at zero distance from the vertical projection of the fault, during the 2004 magnitude **M** 6.6 Niigata, Japan Earthquake (Strasser and Bommer 2009). The NIG019 station in Ojiya, Japan is

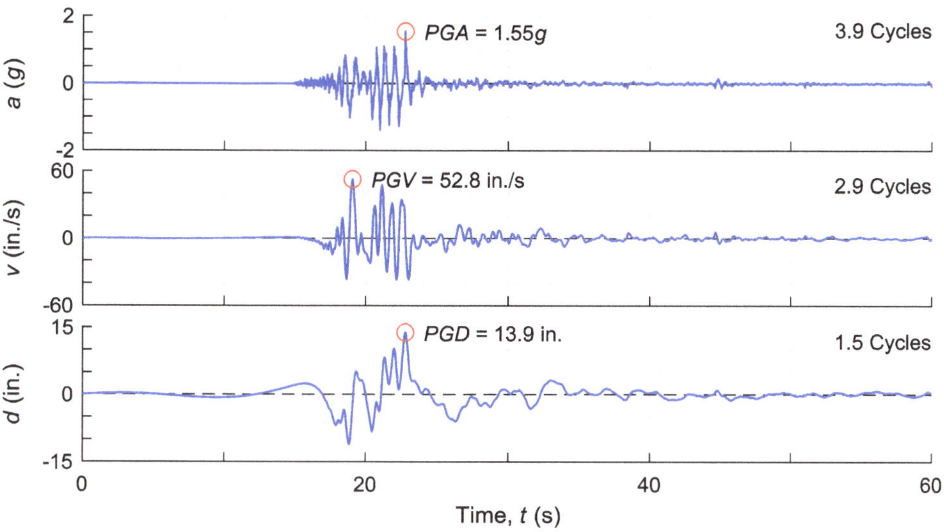

Figure 6-28. Acceleration, velocity, and displacement histories of NIG019 ground motion in maximum-horizontal (RotD100) directions.

owned and maintained by Kyoshin Net. It is situated on Site Class D soil (Strasser and Bommer 2009). The data used in this study were processed by applying baseline correction and low- and high-pass filters. The lowest usable frequency is 0.04 Hz, or the longest usable period is $T_u = 1/0.04 = 25$ s. Figure 6-28 shows acceleration, velocity, and displacement histories of NIG019 ground motion in maximum-horizontal (RotD100) directions. The peak ground acceleration, velocity, and displacement in RotD100 directions are $PGA = 1.55g$, $PGV = 52.8$ in./s, and $PGD = 13.9$ in. The number of cycles in velocity history is 2.9. With $ED = 22.5$ (m/s)^2, $T_{cg} = 0.743$ s, and $PGV_n = 0.58$, NIG019 ground motion is extremely strong, medium frequency, and medium banded.

6.11.1 Response of 3-Story Building to NIG019 Ground Motion

Per Table 5-1, the minimum *ED* required to collapse the 3-story steel SMF building is 23.8 (m/s)^2. The *ED* of NIG019 ground motion is 22.5 (m/s)^2, but it is not centered close to the ultimate period of the building ($T_{cg} = 0.42\ T_{ult}$). Cyclic demand is somewhat high. The NIG019 ground motion is not expected to collapse the 3-story building, but it can damage the building. The response of the building is calculated from both static and dynamic methods of nonlinear analysis described in Chapter 4.

Figure 6-29 shows the demand-capacity equilibrium for the building subjected to NIG019 ground motion. Two demand curves are shown: (1) The magenta curve is the hypothetical demand-curve without any hysteretic damping; it is the 2% damping ADRS; and (2) the black curve is the actual demand-curve for the building generated by the procedure described in Section 4.2. Per static analysis, the CG of the building

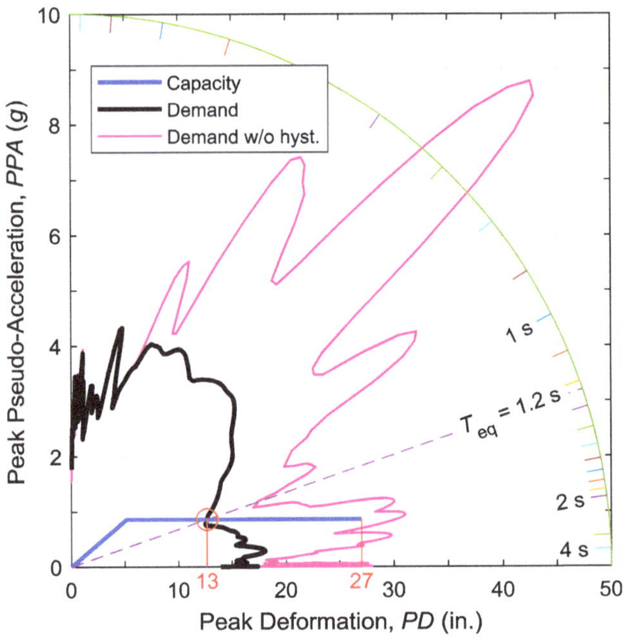

Figure 6-29. Demand-capacity equilibrium for 3-story building subjected to NIG019 ground motion shown in Figure 6-28.

deflects 13 in. The effective period of the building at equilibrium is $T_{eq} = 1.2$ s, and effective damping at equilibrium is $\zeta_{eq} = 12\%$. Using T_{eq} and ζ_{eq}, the equivalent number of cycles, read from cyclic-demand spectra (not shown), is $N_{eq} = 2.9$. Therefore, the CG of the building deflects 13 in. for 2.9 cycles. It was established in Chapter 3 that the CG of the building has the capacity to deflect 27 in. for two cycles. Therefore, the fatigue damage suffered by the building is $(13^2 \times 2.9)/(27^2 \times 2) = 0.32$ (32%) per static analysis. Notice in Figure 6-29 that the high-frequency energy, associated with high *PGA*, is not absorbed by the building. Table 6-19 provides a summary of results from static and dynamic analyses. The results from static analysis are more conservative in this case. If the building had no hysteretic damping (magenta curve in Figure 6-29), it would collapse because the capacity curve stops short of the magenta demand-curve.

Table 6-19. Response of 3-Story Building to NIG019 Ground Motion.

Nonlinear analysis	Roof deflection (in.)	Maximum drift ratio (radian)	Number of cycles	Fatigue damage (%)
Static	21.1	0.0489	2.9	32
Dynamic	19.1	0.0554	2	28

6.11.2 Response of 20-Story Building to NIG019 Ground Motion

Per Table 5-1, the minimum *ED* required to collapse the 20-story steel EBF building is 34.1 $(m/s)^2$. The *ED* of NIG019 ground motion is 22.5 $(m/s)^2$, and it is not centered close to the ultimate period of the building ($T_{cg} = 0.16\ T_{ult}$). Cyclic demand is somewhat high. The NIG019 ground motion is not expected to collapse the 20-story building, but it can damage the building. The response of the building is calculated from both static and dynamic methods of nonlinear analysis described in Chapter 4.

Figure 6-30 shows the demand-capacity equilibrium for the building subjected to NIG019 ground motion. Two demand curves are shown: (1) The magenta curve is the hypothetical demand curve without any hysteretic damping; it is the 2% damping ADRS; and (2) the black curve is the actual demand curve for the building generated by the procedure described in Section 4.2. Per static analysis, the CG of the building deflects 22 in. The effective period of the building at equilibrium is $T_{eq} = 2.6$ s, and effective damping at equilibrium is $\zeta_{eq} = 8.4\%$. Using T_{eq} and ζ_{eq}, the equivalent number of cycles, read from cyclic-demand spectra (not shown), is $N_{eq} = 2.3$. Therefore, the CG of the building deflects 22 in. for 2.3 cycles. It was established in Chapter 3 that the CG of the building has the capacity to deflect 81 in. for one cycle. Therefore, the fatigue damage suffered by the building is $(22^2 \times 2.3)/(81^2 \times 1) = 0.18$ (18%). Notice in Figure 6-30 that the high-frequency energy,

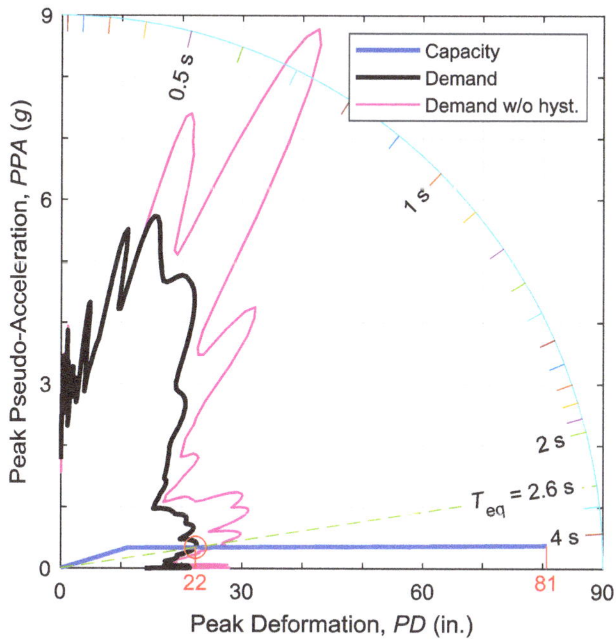

Figure 6-30. Demand-capacity equilibrium for 20-story building subjected to NIG019 ground motion shown in Figure 6-28.

Table 6-20. Response of 20-Story Building to NIG019 Ground Motion.

Nonlinear analysis	Roof deflection (in.)	Maximum drift ratio (radian)	Number of cycles	Fatigue damage (%)
Static	43.4	0.0151	2.3	18
Dynamic	22.4	0.0165	4.1	38

associated with high *PGA*, is not absorbed by the building. Table 6-20 provides a summary of results from static and dynamic analyses. The results from dynamic analysis are more conservative in this case. The fatigue damage from dynamic analysis is 38%. If the building had no hysteretic damping (magenta curve in Figure 6-30), it would deflect 26 in. for 7.9 cycles and suffer 82% fatigue damage.

6.12 Chi-Chi, Taiwan, TCU068 Ground Motion

This ground motion was recorded at the TCU068 station, at zero distance from vertical projection of the fault, during the 1999 magnitude **M** 7.6 Chi-Chi, Taiwan Earthquake. The TCU068 station is owned and maintained by the Central Weather Bureau of Taiwan. Per PEER (2022), the station is situated on Site Class C soil as defined by ASCE 7-22. The data used in this study were processed by applying baseline correction and low- and high-pass filters (PEER 2022). The lowest usable frequency is 0.038 Hz, or the longest usable period is $T_u = 1/0.038 = 26$ s. Figure 6-31 shows acceleration, velocity, and displacement histories of TCU068 ground motion in maximum-horizontal (RotD100) directions. The peak ground acceleration, velocity, and displacement in RotD100 directions are $PGA = 0.578g$, $PGV =$ 75.9 in./s, and $PGD = 128$ in. The number of cycles in velocity history is 1.2. With $ED = 36.2\ (\text{m/s})^2$, $T_{cg} = 6.48$ s, and $PGV_n = 0.45$, TCU068 ground motion is extremely strong, low frequency, and medium to broad banded.

6.12.1 Response of 3-Story Building to TCU068 Ground Motion

Per Table 5-1, the minimum *ED* required to collapse the 3-story steel SMF building is 23.8 $(\text{m/s})^2$. The *ED* of TCU068 ground motion is 36.2 $(\text{m/s})^2$. This ground motion has the energy to collapse the 3-story building, but the energy is centered far away from the ultimate period of the building ($T_{cg} = 3.6\ T_{ult}$). Cyclic demand is low. Therefore, TCU068 ground motion is unlikely to collapse the building despite its enormous energy. The response of the building is calculated from both static and dynamic methods of nonlinear analysis described in Chapter 4.

Figure 6-32 shows the demand-capacity equilibrium for the building subjected to TCU068 ground motion. Two demand curves are shown: (1) The magenta curve is the hypothetical demand curve without any hysteretic damping; it is the 2% damping ADRS; and (2) the black curve is the actual demand curve for the building generated

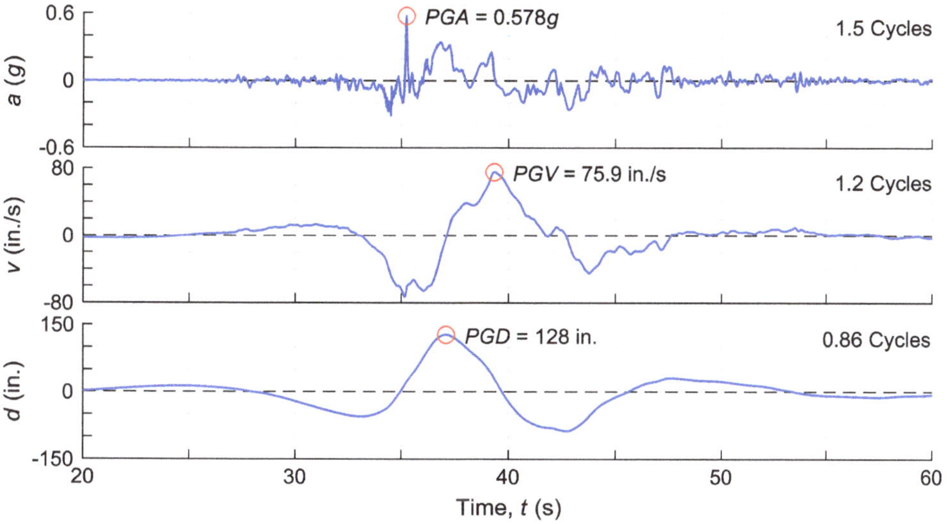

Figure 6-31. Acceleration, velocity, and displacement histories of TCU068 ground motion in maximum-horizontal (RotD100) directions.

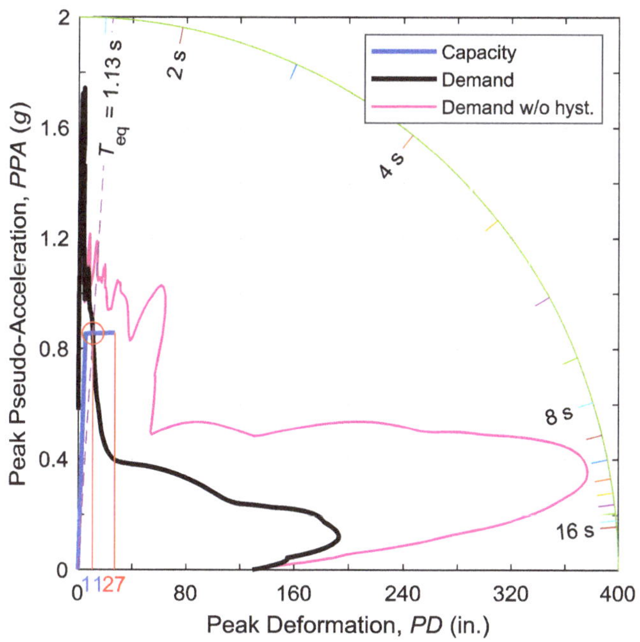

Figure 6-32. Demand-capacity equilibrium for 3-story building subjected to TCU068 ground motion shown in Figure 6-31.

Table 6-21. Response of 3-Story Building to TCU068 Ground Motion.

Nonlinear analysis	Roof deflection (in.)	Maximum drift ratio (radian)	Number of cycles	Fatigue damage (%)
Static	17.7	0.0409	2.1	17
Dynamic	9.7	0.0301	1.9	8

by the procedure described in Section 4.2. Per static analysis, the CG of the building deflects 11 in. The effective period of the building at equilibrium is $T_{eq} = 1.1$ s, and effective damping at equilibrium is $\zeta_{eq} = 8.8\%$. Using T_{eq} and ζ_{eq}, the equivalent number of cycles, read from cyclic-demand spectra (not shown), is $N_{eq} = 2.1$. Therefore, the CG of the building deflects 11 in. for 2.1 cycles. Notice in Figure 6-32 that the low-frequency energy, associated with high *PGD*, is not absorbed by the building. It was established in Chapter 3 that the CG of the building has the capacity to deflect 27 in. for two cycles. Therefore, the fatigue damage suffered by the building is $(11^2 \times 2.1)/(27^2 \times 2) = 0.17$ (17%) per static analysis. Table 6-21 provides a summary of results from static and dynamic analyses. The results from static analysis are more conservative in this case. If the building had no hysteretic damping (magenta curve in Figure 6-32), it would collapse because the capacity curve stops short of the magenta demand-curve.

6.12.2 Response of 20-Story Building to TCU068 Ground Motion

Per Table 5-1, the minimum *ED* required to collapse the 20-story steel EBF building is 34.1 $(m/s)^2$. The *ED* of TCU068 ground motion is 36.2 $(m/s)^2$. It is centered close to the ultimate period of the building ($T_{cg} = 1.38\ T_{ult}$), but it is not narrowly focused ($PGV_n = 0.45$). Cyclic demand is low. TCU068 ground motion is less likely to collapse the building despite its enormous energy. The response of the building is calculated from both static and dynamic methods of nonlinear analysis described in Chapter 4.

Figure 6-33 shows the demand-capacity equilibrium for the building subjected to TCU068 ground motion. Two demand curves are shown: (1) The magenta curve is the hypothetical demand-curve without any hysteretic damping; it is the 2% damping ADRS; and (2) the black curve is the actual demand curve for the building generated by the procedure described in Section 4.2. Per static analysis, the CG of the building deflects 62 in. The effective period of the building at equilibrium is $T_{eq} = 4.2$ s, and effective damping at equilibrium is $\zeta_{eq} = 23\%$. Using T_{eq} and ζ_{eq}, the equivalent number of cycles, read from cyclic-demand spectra (not shown), is $N_{eq} = 0.72$. Therefore, the CG of the building deflects 62 in. for 0.72 cycles. It was established in Chapter 3 that the CG of the building has the capacity to deflect 81 in. for one cycle. Therefore, the fatigue damage suffered by the building is $(62^2 \times 0.72)/(81^2 \times 1) = 0.43$ (43%). Table 6-22 provides a summary of results from static and

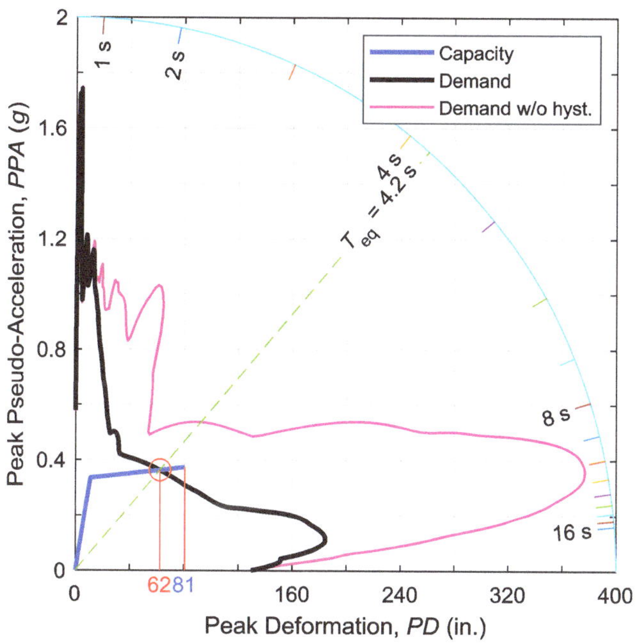

Figure 6-33. Demand-capacity equilibrium for 20-story building subjected to TCU068 ground motion shown in Figure 6-31.

Table 6-22. Response of 20-Story Building to TCU068 Ground Motion.

Nonlinear analysis	Roof deflection (in.)	Maximum drift ratio (radian)	Number of cycles	Fatigue damage (%)
Static	122	0.0423	0.72	43
Dynamic	52	0.0240	0.83	16

dynamic analyses. The results from static analysis are more conservative in this case. Low cyclic-demand is the cause of smaller than expected damage for this ground motion. If the building had no hysteretic damping (magenta curve in Figure 6-33), it would collapse because the capacity curve stops short of the magenta demand curve.

6.13 Damage from Aftershocks

Large earthquakes are usually followed by aftershocks. It takes some time to repair or replace a damaged structure, therefore the structure can suffer additional damage during aftershocks before it is repaired or replaced. It is proposed that all aftershocks that occur within one year of the mainshock be considered an extension of the mainshock. In this section, the additional damage caused by aftershocks is examined for one of the strongest ground motions considered in this chapter.

The maximum damage to the 3-story steel SMF building was caused by the ground motion recorded at TCU084 station during the 1999 magnitude **M** 7.6 Chi-Chi, Taiwan Earthquake. According to Table 6-13, the 3-story building suffers 81% fatigue damage during the TCU084 ground motion per the static analysis. This means that the building is not collapsed but it has used 81% of its fatigue life during the TCU084 ground motion. It is of interest to know if the aftershocks can increase the damage to 100% and cause the building to collapse before it is repaired or replaced.

The 1999 **M** 7.6 Chi-Chi, Taiwan Earthquake produced numerous aftershocks. Within one year of the mainshock, three potentially damaging aftershocks were recorded at the TCU084 station; they are listed in Table 6-23. The four ground motion records listed in Table 6-23 are joined together to generate one long record of 300 s duration, as shown in Figure 6-34. Each individual record has sufficient "quiet" period that ensures that the building comes to rest before it is shaken by the subsequent aftershock.

Table 6-23. Potentially Damaging Ground Motions Recorded at TCU084 Station Within One Year of the 1999 **M** 7.6 Chi-Chi, Taiwan Earthquake.

Description	Magnitude
Mainshock	**M** 7.6
First aftershock	**M** 5.9
Second aftershock	**M** 6.2
Third aftershock	**M** 6.2

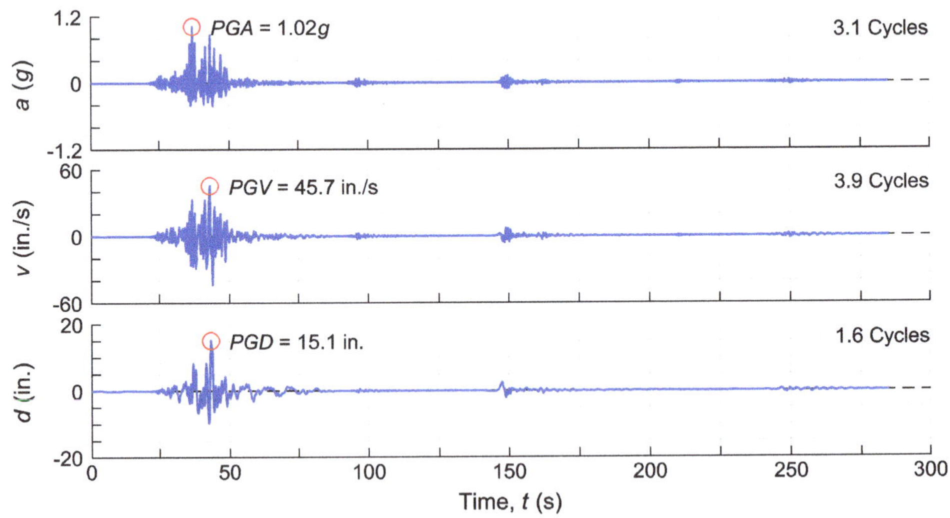

Figure 6-34. Acceleration, velocity, and displacement histories of TCU084 ground motion (mainshock and aftershocks) in maximum-horizontal (RotD100) directions.

The 3-story steel SMF building is re-analyzed by using the entire record shown in Figure 6-34. The damage to the building increases from 81% to 82% due to aftershocks. Such low contribution of aftershocks to cumulative damage is because of the extreme nature of the mainshock ground motion. The effect of aftershocks is expected to be small for the extreme ground motions because they are much stronger than any other ground motions, including aftershocks.

6.14 Summary

1. Both buildings are yielded by each of the 11 strongest ground motions. Therefore, without ductility, both buildings will be collapsed by each of the 11 strongest ground motions.
2. Without hysteretic damping, the 3-story building will be collapsed by 10 out of 11 strongest ground motions and the 20-story building will be collapsed by 6 out of 11 strongest ground motions.
3. Strength alone is not sufficient to prevent collapse by the strongest ground motions. Ductility and hysteresis are also essential to prevent collapse.
4. Cyclic demand can significantly affect the damage caused by strongest ground motions.
5. Ductility and hysteresis reduce both amplitude and cyclic demands imposed by the strongest ground motions.
6. A nonlinear analysis is necessary to consider the effect of ductility and hysteresis on the response of structures.
7. For the 3-story building, the nonlinear-static analysis described in this book is always found to produce more conservative results than the nonlinear-dynamic analysis.
8. For the 20-story building, the nonlinear-static analysis described in this book is found to produce more conservative results seven out of 11 times.
9. The model used for nonlinear-dynamic analysis should be simplified to minimize numerical errors. Dynamic analysis should always be preceded by static analysis.
10. A well-designed structure made of ordinary materials and established technologies can withstand the strongest ground motions.
11. It takes some time to repair or replace a damaged building. During that time, the building can suffer additional damage from aftershocks.
12. It is proposed that all aftershocks that occur within one year of the mainshock should be considered an extension of the mainshock.
13. For the strongest ground motions of interest in this book the effect of aftershocks on cumulative damage is expected to be small because these ground motions are much stronger than any other ground motions, including aftershocks.

14. The most important characteristic of a building is the minimum energy demand *ED* needed to collapse that building. The most important characteristics of a ground motion are its energy demand *ED*, energy centroid T_{cg}, and normalized velocity PGV_n.

References

AFAD (Ministry of Interior, Disaster and Emergency Management Presidency). 2023. *Turkish accelerometric archive v 1.0.* Çankaya/ Ankara, Turkey: AFAD, Earthquake Department.

ASCE. 2022. *Minimum design loads for building and other structures.* ASCE 7-22, Reston, VA: ASCE.

Darragh, R., T. Cao, C. Cramer, V. Graizer, et al. 1994. *Processed CSMIP strong-motion data from the Northridge California Earthquake of 17 January 1994: Release No. 4.* Sacramento, CA: California Strong Motion Instrumentation Program.

Graizer, V., and A. Shakal. 2003. "Directional site amplification effect on Tarzana Hill, California." *American Geophysical Union Fall Meeting 2003,* Abstract ID. S52A-0130.

PEER (Pacific Earthquake Engineering Research Center). 2022. *PEER ground motion database.* Berkeley, CA: University of California. https://ngawest2.berkeley.edu/.

Strasser, F. O., and J. J. Bommer. 2009. "Large-amplitude ground-motion recordings and their interpretations." *Soil Dyn. Earthq. Eng.* 29: 1305–1329.

CHAPTER 7

Collapse-Proof Seismic Design

7.1 Introduction

Collapse-proof structures cannot be collapsed by any of the recorded ground motions. Collapse-proof structures are characterized by high seismic toughness, ST, and high damping. It is proposed that structures with $ST > 5\ (\text{m/s})^2$ be considered extremely tough, and those with $ST < 0.5\,(\text{m/s})^2$ be considered weak. Table 7-1 presents proposed classification of structures based on their seismic toughness, ST. Damping of a structure is determined by its ductility ratio μ and the robustness of its hysteresis loops, defined by β. Structures with high μ can experience large inelastic deformation compared with elastic deformation. Because inelastic damping is significantly higher than elastic damping, higher μ results in higher damping. Structures with higher β have robust hysteresis loops hence higher damping. Seismic design of a collapse-proof structure is discussed in this chapter. The structure is designed to the envelope of 11 strongest ground motions listed in Table 1-2. This is somewhat conservative, but it saves analysis time.

7.2 Envelope Response Spectra

Figure 7-1 shows 10% damping response spectra of 11 strongest ground motions listed in Table 1-2. The 11 response spectra are enclosed within a smooth envelope shown by the thick magenta curve. Envelope response spectra are similarly generated for other values of damping. Figure 7-2 shows envelope response spectra for various damping. Figure 7-3 shows logarithmic ADRS of envelope for various damping. Logarithmic ADRS was previously discussed in Section 1.8. It is simply a tripartite plot turned clockwise by 45 degrees. Envelope cyclic-demand response spectra (CDRS) are similarly generated. Figure 7-4 shows envelope CDRS for various damping.

Table 7-1. Proposed Classification of Structures Based on Their Seismic Toughness, *ST*.

Weak	$ST < 0.5\,(\text{m/s})^2$
Tough	$0.5 \leq ST < 1.5\,(\text{m/s})^2$
Very tough	$1.5 \leq ST < 5\,(\text{m/s})^2$
Extremely tough	$ST \geq 5\,(\text{m/s})^2$

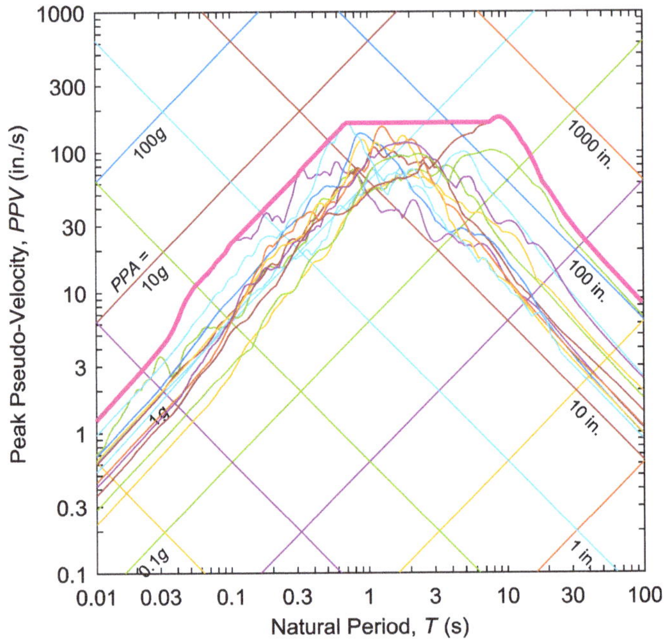

Figure 7-1. Envelope response spectrum of 11 strongest ground motions for 10% of critical damping.

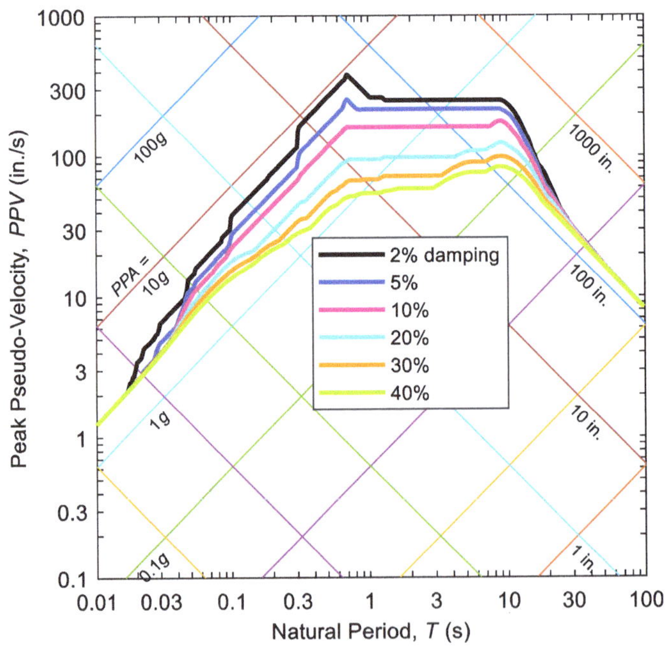

Figure 7-2. Tripartite plots of envelope response spectra for various damping.

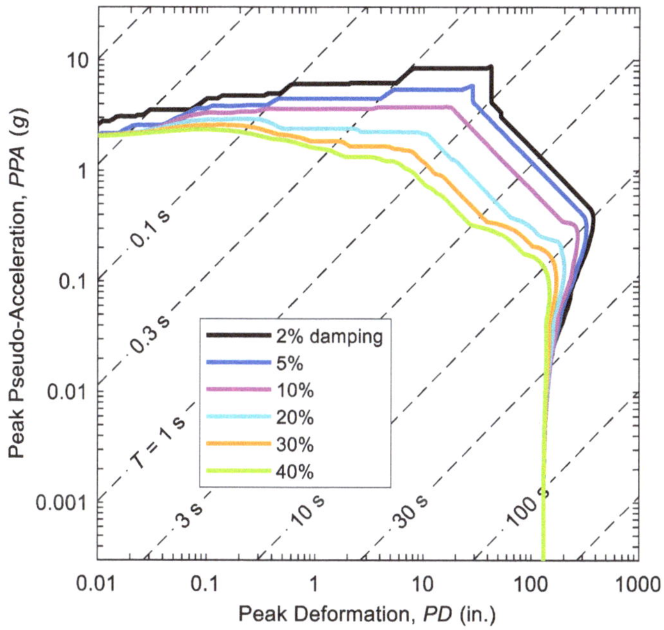

Figure 7-3. Logarithmic ADRS of envelope for various damping.

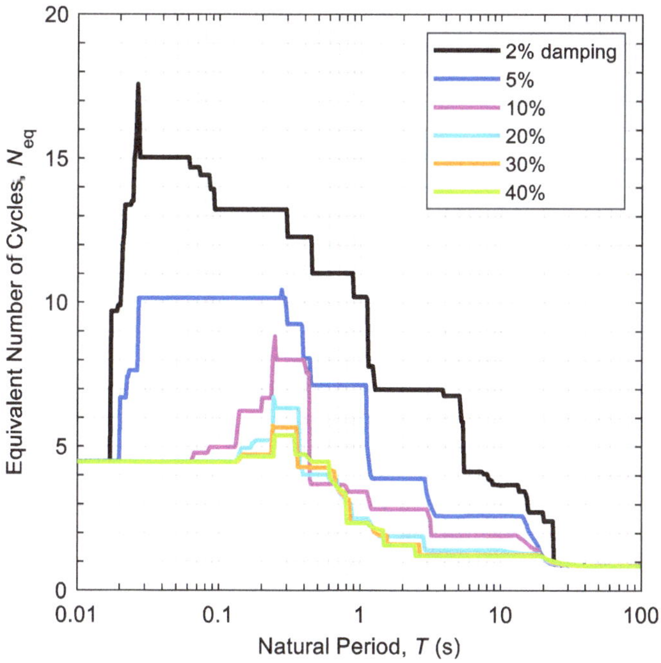

Figure 7-4. Envelope cyclic-demand response spectra for various damping.

7.3 Preliminary Design

Figure 7-5 shows the 3-story steel buckling-restrained braced frame (BRBF) building. A buckling-restrained brace (BRB) can yield both in tension and compression. A normal (concentric) brace buckles in compression before it can yield. But a BRB has an outer encasement that prevents it from buckling while the inner core yields in compression (Watanabe et al. 1988, Tremblay et al. 2006, Kersting et al. 2015, Li et al. 2022). Because of this behavior, a BRB can exhibit a nearly ideal hysteresis loop. A BRB is defined by its axial strength and deformability. The building in Figure 7-5 has the same story heights, bay widths, and seismic weights as the 3-story steel SMF building described in Chapter 3.

In Figure 7-5, the angle of each brace with the horizontal is $\alpha_b = 39°$. Let's assume that the plastic deformability of all braces in the building is $\Delta_b = 4$ in. From geometry, the maximum plastic inter-story drift ratio (Kersting et al. 2015) is

$$\theta_p = \frac{\Delta_b}{h \times \cos \alpha_b} \tag{7-1}$$

where h is the story height. Substituting, $h = 144$ in., $\alpha_b = 39°$, and $\Delta_b = 4$ in. into Equation (7-1), $\theta_p = 0.0356$ radian is obtained. Ignoring elastic deflection, the maximum deformability of the building is

$$D_{ult} \approx \theta_p \times H_{CG} \tag{7-2}$$

where H_{CG} is the height of the CG. Substituting, $\theta_p = 0.0356$ radian, and $H_{CG} = 259$ in. into Equation (7-2), $D_{ult} \approx 9.2$ in. is obtained. The roof deflection is $\approx \theta_p \times H = 0.0356 \times 432 = 15$ in. Hence, 15 in. clearance is needed all around the building to allow it to deform freely.

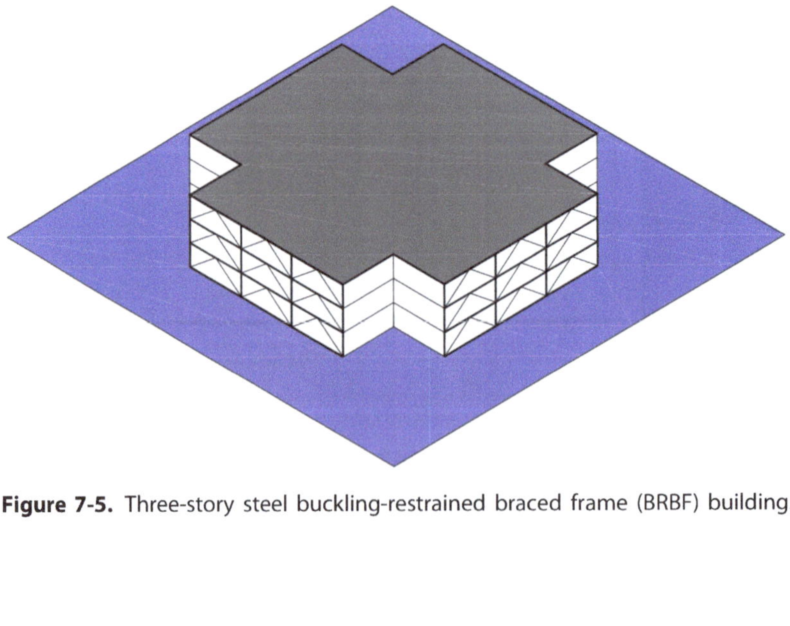

Figure 7-5. Three-story steel buckling-restrained braced frame (BRBF) building.

For preliminary sizing, let's assume that the damping of the building at ultimate deflection is 20% of critical. Corresponding to 9.2 in. deflection and 20% damping, the peak pseudo-acceleration per Figure 7-3 is PPA = 2.22g. The required lateral strength of the building is

$$F = M \times PPA \tag{7-3}$$

Substituting, M = 3,300 kip/g, and PPA = 2.22g into Equation (7-3), F = 7,340 kip is obtained. This is the total base shear resisted by 12 braces in the first story. From geometry, the required axial strength of each brace in the first story is

$$P_{b1} = F/(12 \times \cos\alpha_b) \tag{7-4}$$

Substituting, F = 7,340 kip, and α_b = 39 degrees into Equation (7-4), $P_{b1} = 783$ kip is obtained. Assuming force in upper stories to be proportional to the mass above those stories, the required brace force in second and third stories is P_{b2} = 783 × (3/5) = 470 kip, and P_{b3} = 783 × (1/5) = 157 kip, respectively.

7.4 Capacity and Damping Curves

The BRB manufacturer/supplier is provided with the required strength and deformability of braces and requested to provide force–deformation relationships for braces based on tests. The manufacturer suggests braces with yield strength of 821, 479, and 171 kip for first, second, and third stories, respectively. The braces have been tested per the protocol in Table C-K3.1 of AISC 341-16. It is assumed that the braces can deform 4 in. and develop specified strength for two cycles. Beam and column sections are selected to remain elastic during fully yielded and strain-hardened braces. The first-mode elastic period of the building is calculated to be T_e = 0.21 s. Elastic damping is assumed to be 2% of critical. The capacity curve of the building is generated by following a similar procedure as described in Chapter 3. It is shown in Figure 7-6. The elastic deflection is D_e = 0.98 in., and the ultimate deflection is D_{ult} = 10.2 in. The ductility ratio of the building is μ = 10.2/0.98 = 10.4. The seismic toughness of the building is ST = 5.86 (m/s)^2.

As a sanity check, approximate value of ST is obtained by adding the toughness of all structural elements that can yield and dividing that by the mass of structure. There are 12 braces each of strength 821, 479, and 171 kip, in first, second, and third stories, respectively. The deformability of each brace is 4 in. The total work done by fully yielded braces is 12 × (821 + 479 + 171) × 4 = 70,608 kip-in. Dividing this by the mass of the building gives $ST \approx$ 70,608/(3,300/386) = 8,259 (in./s)^2 = 5.33 (m/s)^2. This is only 9% smaller than the actual value of 5.86 (m/s)^2.

Figure 7-7 shows the damping curve for the building generated by the method described in the appendix. Note that α = 0.009 and β = 0.7 are assumed to generate

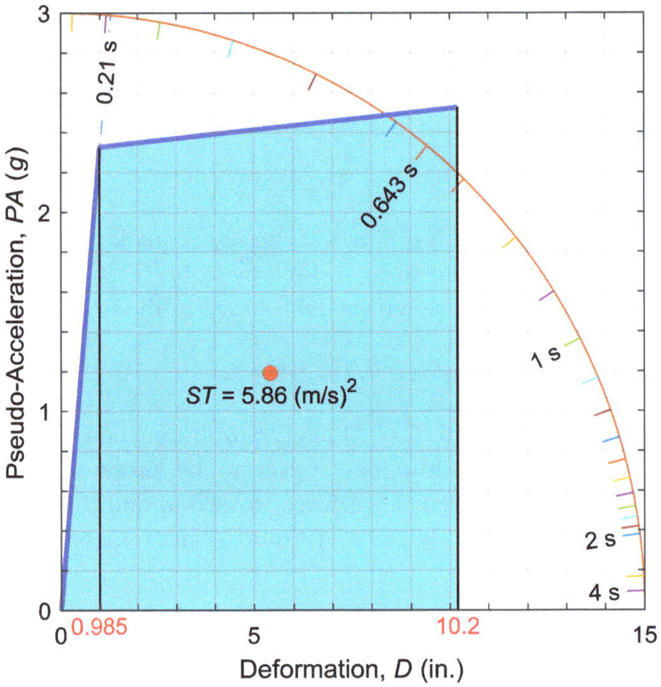

Figure 7-6. Capacity curve of the building shown in Figure 7-5.

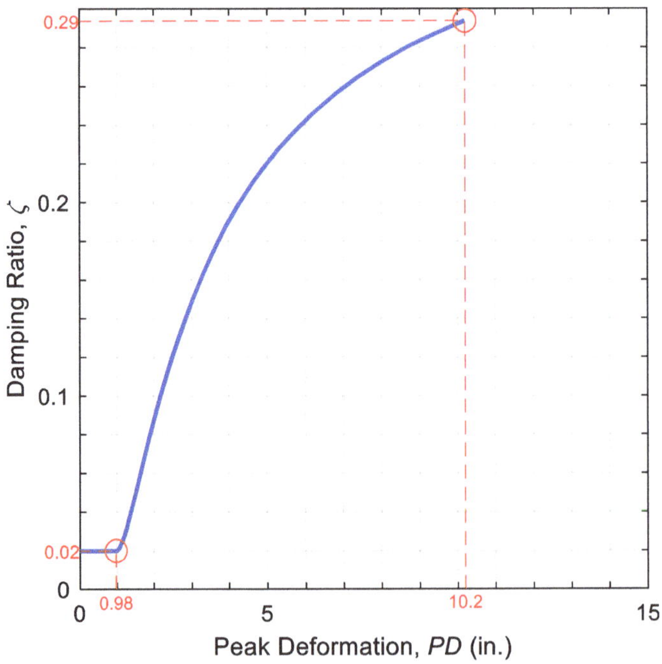

Figure 7-7. Damping curve of the building shown in Figure 7-5.

the damping curve. The building achieves a maximum damping of 29% of critical because of its very high ductility ratio of $\mu = 10.4$.

7.5 Soil–Structure Interaction

The soil below the foundation of the structure can deform and radiate energy away from the structure (Veletsos 1977, Gazetas 1991). Soil–structure interaction (SSI) increases the elastic period of the system and generally increases its initial (elastic) damping. The SSI depends on the size of the foundation, flexibility of the soil below the foundation (measured by its shear modulus), and the elastic period of the structure. For this building the SSI causes the elastic period to increase from 0.21 to 0.29 s and the elastic damping to increase from 2% to 14% of critical. The capacity and damping curves are regenerated using new values of elastic period and damping. Figures 7-8 and 7-9 show the capacity and damping curves, respectively, after considering SSI. Soil–structure interaction increases the elastic deflection from 0.98 to 1.87 in. and ultimate deflection from 10.2 to 11.1 in. As a result, the seismic toughness increases from $ST = 5.86$ to 6.11 $(m/s)^2$, but the ductility ratio reduces from $\mu = 10.4$ to 5.9. Owing to significant reduction in μ, the maximum damping of the building reduces from 29% to 28% of critical (Figure 7-9), despite a significant increase in initial (elastic) damping from 2% to 14% of critical.

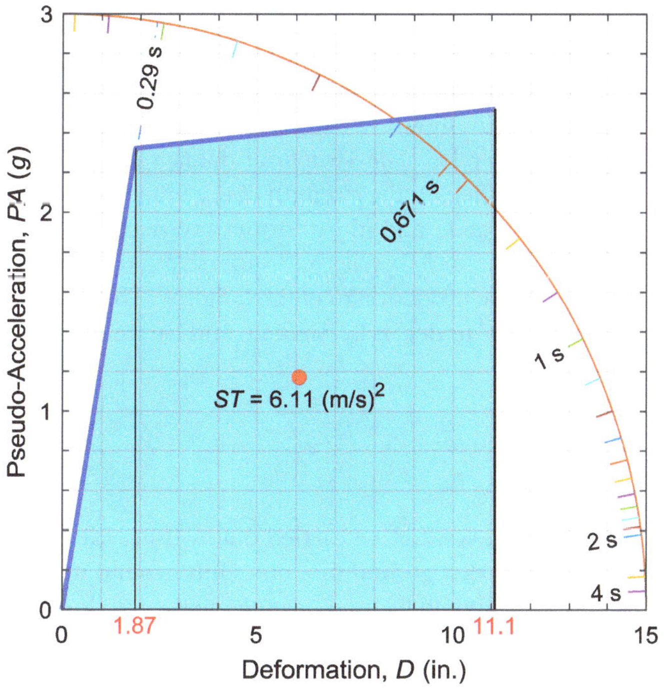

Figure 7-8. Capacity curve of the building considering soil-structure interaction.

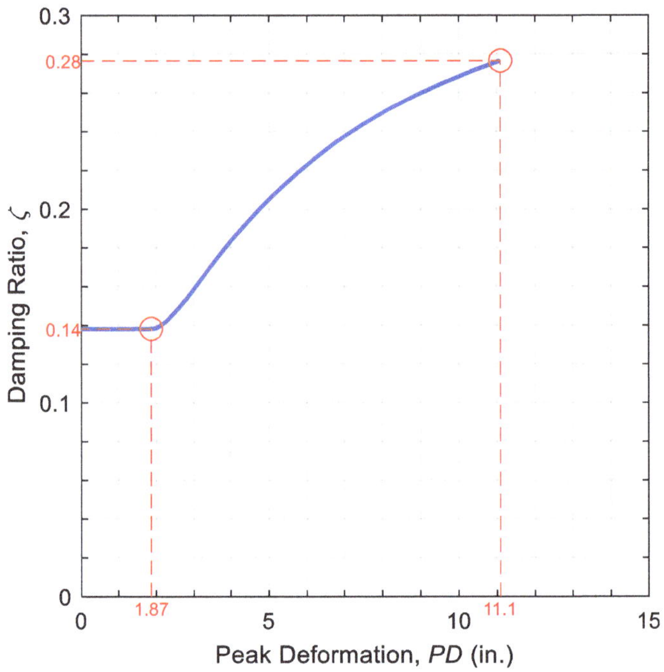

Figure 7-9. Damping curve of the building considering soil-structure interaction.

7.6 Final Design

By using the method described in Chapter 3, the building is analyzed only once using the envelope response spectra shown in Figure 7-3. The CG of the building is found to deflect 4.36 in. for 4.1 cycles. The structural damage to the building is 32%. To make the design more efficient, the strength of the braces is reduced to 547, 342, and 137 kip for first, second, and third stories, respectively. The building is reanalyzed. It deflects 7.7 in. for 3.6 cycles and suffers 95% damage. This concludes selection of BRBs for collapse-proof design of the building. The rest of the building is designed to remain elastic under fully yielded and strain hardened BRBs per AISC 341-16 (AISC 2016).

7.7 Summary

1. Collapse-proof seismic design can be carried out by using the envelope response spectra of strongest recorded ground motions. Only ground motions with energy demand $ED > 10\ (\text{m/s})^2$ should be used to generate the envelope, because weaker ground motions do not have the energy to collapse well-designed structures in high seismic areas.

2. Strength of a structure should be determined from its deformability and damping, rather than its elastic period and 5% damping. Deformability and damping of a structure should be confirmed through a nonlinear analysis.
3. SSI increases the deformability hence seismic toughness ST of the building. SSI generally increases the initial (elastic) damping of the building, but it reduces its ductility ratio μ. Despite increase in initial damping, the overall damping can still reduce owing to the decrease in μ.
4. Collapse-proof structures are different from collapse-resistant structures. Collapse-resistant structures are designed to withstand ground motions in codes and regulations. These ground motions have a certain chance of exceedance at specific sites. Collapse-proof structures are designed to withstand the strongest ground motions ever recorded. These ground motions have a negligible chance of exceedance at specific sites.

References

AISC (American Institute of Steel Construction). 2016. *Seismic provisions for structural steel buildings*. AISC 341-16. Chicago: AISC.

Li, C-H., Z. Vidmar, B. Saxey, M. Reynolds, et al. 2022. "A procedure for assessing low-cycle fatigue life of buckling-restrained braces," *J. Struct. Eng.* 148 (2): 2022;148:04021259. https://doi.org/10.1061/(ASCE)ST.1943-541X.0003237.

Gazetas, G. 1991. "Formulas and charts for impedances of surface and embedded foundations." *J. Geotech. Eng.* 117 (9): 1363–1381.

Kersting, R. A., L. A. Fahnestock, and W. A. López. 2015. "Seismic design of steel buckling-restrained braced frames—A guide for practicing engineers." NEHRP seismic design Tech. Brief No. 11, *NIST GCR 15-917-34*. Gaithersburg, MD: National Institute of Standards and Technology.

Tremblay, R., P. Bolduc, R. Neville, and R. DeVall. 2006. "Seismic testing and performance of buckling restrained bracing systems." *Can. J. Civil Eng.* 33: 183–198. https://doi.org/10.1139/105-103.

Veletsos, A. S. 1977. "Dynamics of structure-foundation systems." In *Structural and geotechnical mechanics*, a volume honoring N. M. Newmark. Edited by W. J. Hall. Englewood Cliffs, NJ: Prentice Hall, 333–361.

Watanabe, A., Y. Hitomi, E. Saeki, A. Wada, et al. 1988. "Properties of brace encased in buckling-restraining concrete and steel tube." In *Proc., 9th World Conference on Earthquake Engineering*, Vol. 4. Tokyo: Japan Association for Earthquake Disaster Prevention.

CHAPTER 8

R, C_d, and Ω_0 Factors

8.1 Introduction

Per ASCE 7-22, ordinary structures should not be collapsed by the MCE_R ground motions as defined by ASCE 7-22 (ASCE 2022). MCE_R ground motions are consensus based. They are not the maximum recorded or the maximum possible. They can be exceeded but their mean return period (MRP) of exceedance is different in different parts of the country (Malhotra 2005). The design ground motions in ASCE 7-22 are two-thirds of the MCE_R ground motions. Routine seismic design does not require a nonlinear analysis; it is based on an elastic analysis. The peak pseudo-acceleration PPA (spectral acceleration) is read from the 5% damping design (2/3 MCE_R) response spectrum using the elastic period of the structure. The design strength is obtained by multiplying the PPA by seismic mass of the structure and dividing by an R factor of up to 8. The R factor is in addition to the two-thirds factor applied on the MCE_R spectrum to obtain the design spectrum. Therefore, the elastic strength-demand is reduced by $1.5R$ to obtain the design strength. The $1.5R$ reduction is supposed to account for the following in an ad hoc manner:

1. Due to yielding, the actual period of the structure, during MCE_R ground motion, is longer than its elastic period.
2. Due to hysteresis, the actual damping of the structure, during MCE_R ground motion, is higher than 5% of critical.
3. The actual strength of a fully yielded, strain-hardened structure is higher than its design strength. The ratio between the actual and design strengths is the over-strength factor Ω_0.

The deflections obtained from the elastic analysis are amplified by $1.5C_d$ factor to account for the contribution of plastic deflections. This chapter explores if the R, C_d, and Ω_0 factors can be rationally determined for use in collapse-proof seismic design. Examples of the 3-story steel SMF building and the 20-story steel EBF building are used for this purpose. To keep the discussions simple, the effect of cyclic demand is not considered in this chapter.

8.2 Three-Story Steel Special Moment Frame Building

The Ω_0, C_d, and R factors for the 3-story steel SMF building described in Chapter 3 are determined in this section.

8.2.1 Overstrength Factor Ω_0

For ASTM A992 steel, the design bending stress is $\phi F_y = 0.9 \times 50 = 45$ ksi per AISC 341-16 (AISC 2016). The design strength of the building is the maximum lateral force that can be applied without exceeding the design stress anywhere in the building. A lateral force of $F = 1{,}000$ kip is distributed among various floors of the building by using Equation (12.8-12) in ASCE 7-22. These forces are $F_1 = 353$ kip at Level 1, $F_2 = 445$ kip at Level 2, and $F_3 = 202$ kip at Level 3 (roof). Forces F_1, F_2, and F_3 are applied to the building and bending moments calculated in beams at various levels using SAP 2000 (CSI 2011). The maximum bending moment of 11,910 kip-in. occurs in the first story beams in the outer bays of the building. (For this building, the moments in the SMF due to the gravity loads are negligible.) The tensile stress in the beam flange induced by 11,910 kip-in. bending moment is 36.25 ksi. This is 0.8055 times the design stress of 45 ksi. Therefore, the design lateral strength of the building is $1{,}000/0.8055 = 1{,}241$ kip. The actual strength of the fully yielded, strain-hardened building is 3,180 kip, per Figure 3-5. Therefore, the overstrength factor for this building is $\Omega_0 = 3{,}180/1{,}241 = 2.56$. Notice that the effect of P-Δ on lateral strength is not considered in the calculation of the overstrength factor because P-Δ does not reduce the stresses in the structural members; it reduces the lateral force that can be applied on the building. Overstrength factor $\Omega_0 = 2.56$ is comparable to the value of 3 for steel SMF per Table 12.2-1 of ASCE 7-22. But, Ω_0 is specific to the building; it could change significantly if the SMF also resisted the gravity loads or if the beams at various levels of the building had different cross sections.

8.2.2 Deflection Amplification Factor C_d

The elastic deflection of the building upon reaching its design lateral strength of 1,241 kip is $1{,}241/k$, where k is the elastic stiffness of the building. Substituting $k = 559$ kip/in. from Section 3.3, the elastic design deflection is $1{,}241/559 = 2.22$ in. The ratio between the ultimate deflection and elastic design deflection is $1.5C_d = 27/2.22 = 12$. This gives $C_d = 8$. This is greater than the value of 5.5 for steel SMF per Table 12.2-1 of ASCE 7-22. Like Ω_0, C_d is specific to the building; it could change significantly if the SMF also resisted the gravity loads or if the beams at various levels of the building had different cross sections.

8.2.3 Response Modification Factor R

Returning to Figures 3-7 and 3-11 in Chapter 3, the effective period and damping of the 3-story steel SMF building at collapse are 1.79 s and 23% of critical, respectively. The ultimate deflection of the building is $D_{ult} = 27$ in. Per the analysis in Section 4.3, the building is not collapsed by the TAPS ground motion described in Chapter 1. Figure 4-1 shows the ADRS of TAPS ground motion for various values of damping. The peak deformation corresponding to 1.79 s period and 23% damping is $PD = 14.8$ in. This is 1/1.83 times the ultimate deflection of $D_{ult} = 27$ in. The safety factor against collapse is SF = 1.83, or the building collapses when subjected to 1.83 times the TAPS ground motion. Figure 8-1 shows 5% and 23% damping ADRS of 1.83 × TAPS ground motion superimposed on the capacity curve of the 3-story steel SMF building described in Chapter 3. Notice that the capacity curve barely intersects the 23% damping ADRS, implying that the building is just collapsed by 1.83 × TAPS ground motion.

The elastic period of the building is $T_e = 0.777$ s per Chapter 3. The pseudo-acceleration corresponding to elastic period and 5% damping is $1.71g$ per Figure 8-1. The elastic strength-demand for the building is $1.71W = 5{,}652$ kip, where $W = 3{,}300$ kip is the seismic weight of the building. The design lateral strength of the building is 1,241 kip as determined in Section 8.2.1. Therefore,

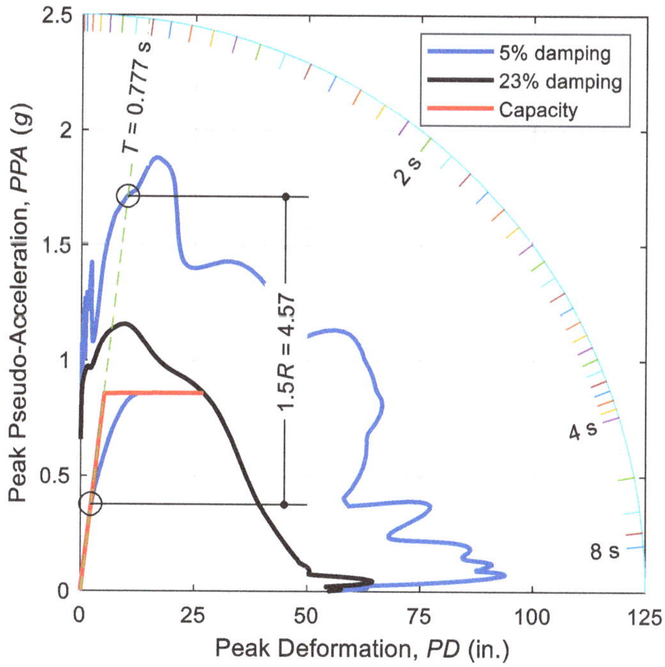

Figure 8-1. Five percent and 23% damping acceleration–deformation response spectra of 1.83 × TAPS ground motion superimposed on the capacity curve of 3-story steel special moment frame building described in Chapter 3.

Table 8-1. Response Modification Factor *R* for the 3-Story Steel SMF Building Subjected to Various Ground Motions.

No.	Earthquake	Recording Station	SF	*R*
1	1994 **M** 6.7 Northridge, CA	Tarzana Cedar Hill	3.0	10
2	1989 **M** 6.9 Loma Prieta, CA	LGPC	1.8	7.1
3	1999 **M** 7.6 Chi-Chi, Taiwan	TCU065	2.0	4.8
4	1994 **M** 6.7 Northridge, CA	Sylmar Converter	1.7	6.4
5	1999 **M** 7.6 Chi-Chi, Taiwan	CHY080	1.8	8.8
6	1995 **M** 6.8 Kobe, Japan	Takatori	1.2	3.5
7	1999 **M** 7.6 Chi-Chi, Taiwan	TCU084	1.3	5.6
8	2023 **M** 7.8 Nurdagi, Türkiye	TK3123	1.1	2.7
9	1999 **M** 7.6 Chi-Chi, Taiwan	TCU052	1.4	2.4
10	2004 **M** 6.6 Niigata, Japan	NIG019	2.2	14
11	1999 **M** 7.6 Chi-Chi, Taiwan	TCU068	1.9	3.3

$1.5R = 5{,}652/1{,}241 = 4.57$ or $R = 3$. This is much smaller than the value of 8 for steel SMF per Table 12.2-1 of ASCE 7-22.

The response modification factor *R* is similarly computed for all 11 ground motions in Table 1-2. The first three columns in Table 8-1 identify the ground motions. The fourth column lists the safety factor SF against collapse. This is the factor by which the ground motion is multiplied to barely collapse the building. The last column in Table 8-1 lists the calculated values of the *R* factor. Note that the *R* factor is different for different ground motions; it ranges from 2.4 to 10. Ground motions are composed of many frequencies. For a ductile structure, frequencies that excite the elastic response are very different from those that collapse the structure. Therefore, the *R* factor is very sensitive to the frequency composition of the ground motion. A fixed value of *R* for all ground motions is unreasonable.

The 3-story steel SMF building is also subjected to the envelope ground motion described in Chapter 7. The safety factor against collapse is SF = 1.04, meaning that the building just avoids collapse by the envelope ground motion. Figure 8-2 shows 5% and 23% damping ADRS of 1.04 × envelope ground motion superimposed on the capacity curve of the 3-story steel SMF building. The ratio between the elastic strength demand and the design strength is $1.5R = 13.5$. The *R* factor for the 3-story steel SMF building subjected to the envelope ground motion is $R = 9$.

Note that elastic analysis does not capture the true strength, deformability, and damping of a structure. The Ω_0, C_d, and *R* factors are used to mitigate the shortcoming of elastic analysis in an ad hoc manner. The first column in Table 8-2 lists the elastic strength, deformability, and damping of the 3-story SMF building. With such low values of strength, deformability, and damping, the building cannot withstand the TAPS ground motion. The safety factor SF against collapse is listed in the last row of first column. A safety factor less than 1 implies that the building collapses during the TAPS ground motion. The second column in Table 8-2 lists the true (inelastic)

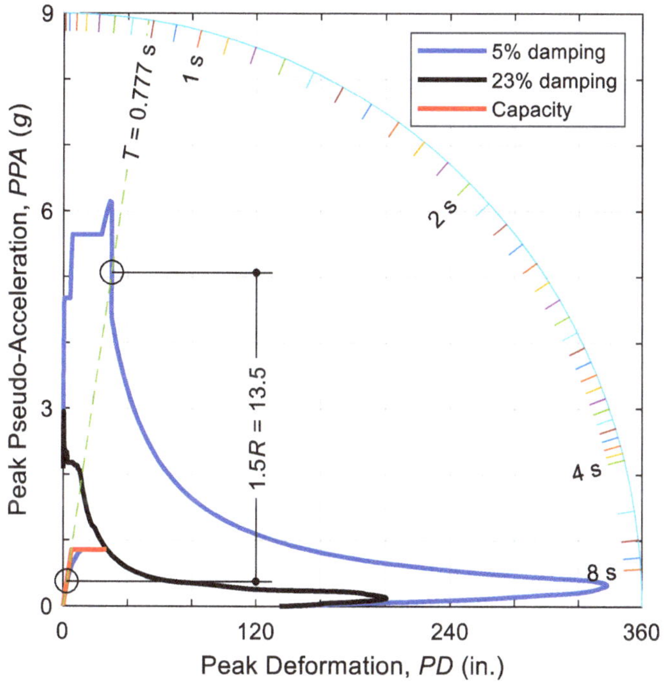

Figure 8-2. Five percent and 23% damping acceleration–deformation response spectra of 1.04 × envelope ground motion described in Chapter 7 superimposed on the capacity curve of the 3-story steel SMF building described in Chapter 3.

Table 8-2. Elastic and Inelastic Strength, Deformability, Damping, and Safety Factor Against Collapse of 3-Story Steel SMF Building Subjected to TAPS Ground Motion.

	Elastic	Inelastic	
Strength	1,241 kip	3,180 kip	$\Omega_0 = 3{,}180/1{,}241 = 2.56$
Deformability	2.22 in.	27 in.	$1.5C_d = 27/2.22 = 12$
Damping	5% of critical	23% of critical	–
Safety factor against collapse	0.40	1.83	$1.5R = 1.83/0.40 = 4.57$

strength, deformability and damping of the 3-story SMF building. Such high values of strength, deformability, and damping allow the building to withstand the TAPS ground motion. The SF against collapse is listed in the last row of the second column. A safety factor greater than 1 implies that the building is not collapsed by the TAPS ground motion. The ratio between inelastic and elastic strengths is the overstrength factor Ω_0. The ratio between inelastic and elastic deformability is $1.5C_d$. The ratio between inelastic and elastic safety factors is $1.5R$. These values are listed in the last column of Table 8-2.

8.3 Twenty-Story Steel Eccentrically Braced Frame Building

The Ω_0, C_d, and R factors for the 20-story steel EBF building, described in Chapter 3, are determined in this section.

8.3.1 Overstrength Factor Ω_0

For ASTM A992 steel, the design shear stress is $\phi 0.6F_y = 0.9 \times 0.6 \times 50 = 27$ ksi per AISC 341-16 (AISC 2016). The design strength of the building is the maximum lateral force that can be applied without exceeding the design stress anywhere in the building. A lateral force of $F = 1{,}000$ kip is distributed among various floors of the building by using Equation (12.8-12) in ASCE 7-22. These forces are applied to the building and shear stresses are calculated in link-beams at various levels using SAP 2000 (CSI 2011). The maximum shear stress of 7.46 ksi occurs in the innermost link-beams at the ninth level of the building. The maximum shear stress is 0.2764 times the design shear stress of 27 ksi. Therefore, the design lateral strength of the building is 1,000/0.2764 = 3,620 kip. The actual strength of the fully yielded, strain-hardened building is 10,970 kip, per Figure 3-18. Therefore, the overstrength factor for this building is $\Omega_0 = 10{,}970/3{,}620 = 3$. Overstrength $\Omega_0 = 3$ is greater than the value of 2 for steel EBF per Table 12.2-1 of ASCE 7-22. Note that Ω_0 is specific to the building. Ω_0 could change significantly if the link-beams at various levels of the building had different cross sections.

8.3.2 Deflection Amplification Factor C_d

The elastic design deflection of the building upon reaching its design lateral strength of 3,620 kip is 3,620/k, where k is the elastic stiffness of the building. Substituting $k = 792$ kip/in. from Section 3.7, the elastic design deflection is 3,620/792 = 4.57 in. The ratio between the ultimate deflection and elastic design deflection is $1.5C_d = 80.7/4.57 = 17.6$. This gives $C_d = 12$. This is three times the value of 4 for steel EBF per Table 12.2-1 of ASCE 7-22 (2022). Like Ω_0, C_d is specific to the building; it could change significantly if the link-beams at various levels had different cross sections.

8.3.3 Response Modification Factor R

Returning to Figures 3-20 and 3-22, the effective period and damping of the 20-story steel EBF building at collapse are 4.7 s and 25% of critical, respectively. The ultimate deflection of the building is $D_{ult} = 80.7$ in. Per the analysis in Section 4.7, the building is not collapsed by the TAPS ground motion described in Chapter 1. Figure 4-1 shows the ADRS of TAPS ground motion for various values of damping.

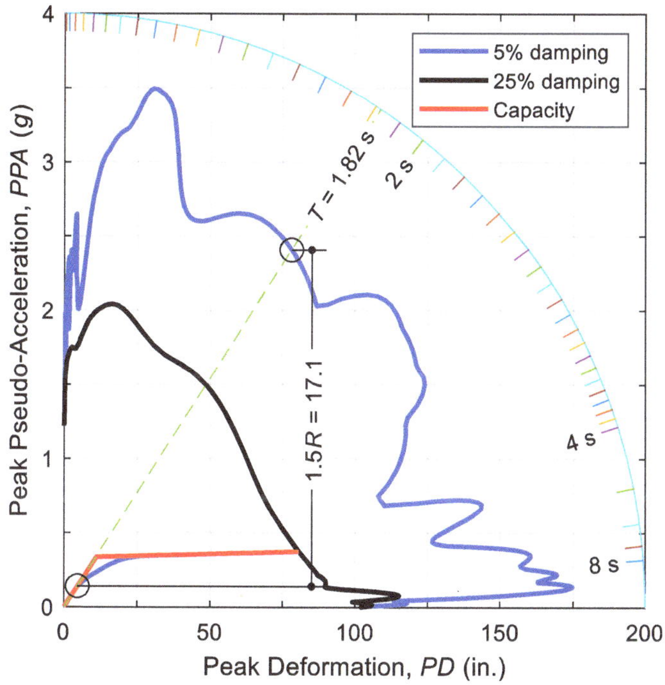

Figure 8-3. Five percent and 23% damping acceleration-deformation response spectra of 3.4 × TAPS ground motion superimposed on the capacity curve of 20-story steel eccentrically braced frame building described in Chapter 3.

The peak deformation corresponding to 4.7 s period and 25% damping is $PD = 23.7$ in. This is 1/3.4 times the ultimate deflection of $D_{ult} = 80.7$ in. The safety factor against collapse is SF = 3.4, or the building is collapsed by 3.4 times the TAPS ground motion. Figure 8-3 shows 5% and 23% damping ADRS of 3.4 × TAPS ground motion superimposed on the capacity curve of 20-story steel EBF building described in Chapter 3. Notice that the capacity curve barely intersects the 25% damping ADRS, implying that the building is just collapsed by 3.4 × TAPS ground motion.

The elastic period of the building is $T_e = 1.82$ s per Chapter 3. The pseudo-acceleration corresponding to elastic period and 5% damping is $2.41g$ per Figure 8-3. The elastic strength-demand for the building is $2.41W$ = 61,750 kip, where W = 25,640 kip is the seismic weight of the building. The design lateral strength of the building is 3,620 kip as determined in Section 8.3.1. Therefore, $1.5R$ = 61,750/3,620 = 17 or R = 11. This is greater than the value of 8 for steel EBF per Table 12.2-1 of ASCE 7-22.

The response modification factor R is similarly computed for all 11 ground motions in Table 1-2. The first three columns in Table 8-3 identify the ground motions. The fourth column lists the safety factor SF against collapse. This is the factor by which the ground motion is multiplied to barely collapse the 20-story steel

Table 8-3. Response Modification Factor *R* for the 20-Story Steel EBF Building Subjected to Various Ground Motions.

No.	Earthquake	Recording Station	SF	*R*
1	1994 **M** 6.7 Northridge, CA	Tarzana Cedar Hill	6.3	12
2	1989 **M** 6.9 Loma Prieta, CA	LGPC	2.5	7.7
3	1999 **M** 7.6 Chi-Chi, Taiwan	TCU065	1.7	7.3
4	1994 **M** 6.7 Northridge, CA	Sylmar Converter	3.4	17
5	1999 **M** 7.6 Chi-Chi, Taiwan	CHY080	5.1	17
6	1995 **M** 6.8 Kobe, Japan	Takatori	3.9	25
7	1999 **M** 7.6 Chi-Chi, Taiwan	TCU084	5.1	37
8	2023 **M** 7.8 Nurdagi, Türkiye	TK3123	2.6	15
9	1999 **M** 7.6 Chi-Chi, Taiwan	TCU052	1.7	8.2
10	2004 **M** 6.6 Niigata, Japan	NIG019	5.4	20
11	1999 **M** 7.6 Chi-Chi, Taiwan	TCU068	1.1	4.2

EBF building. The last column in Table 8-3 lists the calculated values of the R factor. Like the 3-story building, the R factor depends on the ground motion. It varies from 4.2 to 37. The R factor is very sensitive to the frequency composition of ground motion. A fixed value of R for all ground motions is unreasonable.

The 20-story steel EBF building is also subjected to the envelope ground motion described in Chapter 7. The safety factor against collapse is SF = 1.1, meaning that the building collapses when subjected to 1.1 times the envelope ground motion. Figure 8-4 shows 5% and 23% damping ADRS of 1.1 × envelope ground motion superimposed on the capacity curve of the 20-story steel EBF building. The ratio between the elastic strength demand and the design strength is $1.5R = 15.5$. The R factor for the 20-story steel EBF building subjected to envelope ground motion is $R = 15.5/1.5 = 10$.

The first column in Table 8-4 lists the elastic strength, deformability, and damping of the 20-story EBF building. With such low values of strength, deformability, and damping, the building cannot withstand the TAPS ground motion. The safety factor SF against collapse is listed in the last row of the first column. A safety factor less than 1 implies that the building collapses during the TAPS ground motion. The second column in Table 8-4 lists the true (inelastic) strength, deformability, and damping of the 20-story EBF building. Such high values of strength, deformability, and damping allow the building to withstand the TAPS ground motion. Thc safcty factor SF against collapse is listed in the last row of the second column. A safety factor greater than 1 implies that the building is not collapsed by the TAPS ground motion. The ratio between inelastic and elastic strengths is the overstrength factor Ω_0. The ratio between inelastic and elastic deformability is $1.5C_d$. The ratio between inelastic and elastic safety factors is $1.5R$. These values are listed in the last column of Table 8-4.

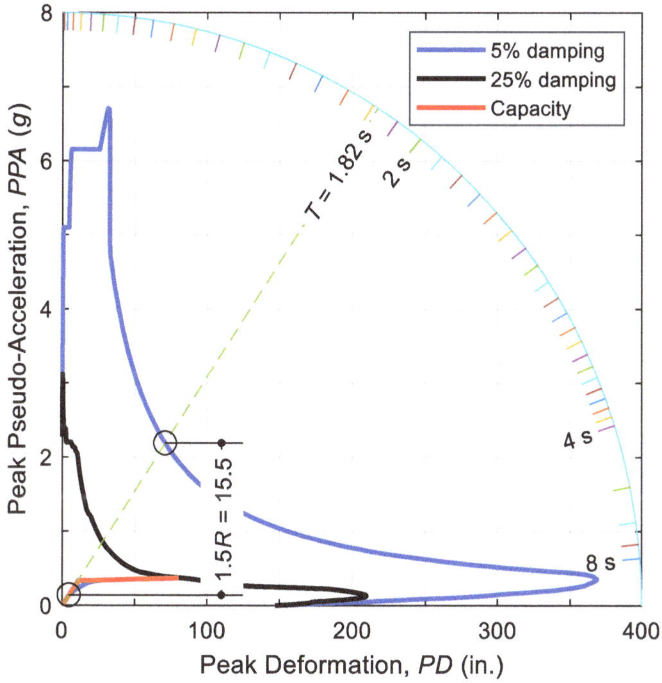

Figure 8-4. Five percent and 23% damping acceleration-deformation response spectra of 1.1 × envelope ground motion described in Chapter 7 superimposed on the capacity curve of the 20-story steel EBF building described in Chapter 3.

Table 8-4. Elastic and Inelastic Strength, Deformability, Damping, and Safety Factor Against Collapse of the 20-Story Steel EBF Building Subjected to TAPS Ground Motion.

	Elastic	Inelastic	
Strength	3,620 kip	10,970 kip	$\Omega_0 = 10{,}970/3{,}620 = 3$
Deformability	4.57 in.	80.7 in.	$1.5C_d = 80.7/4.57 = 17.6$
Damping	5% of critical	23% of critical	–
Safety factor against collapse	0.20	3.4	$1.5R = 3.4/0.20 = 17$

8.4 Summary

1. The overstrength factor Ω_0 and the deflection amplification factor C_d can vary from one building to another even if the buildings have the same lateral load resisting system.

2. The response modification factor R is very sensitive to the frequency composition of the ground motion. It is unreasonable to assume the same R for all ground motions.
3. Elastic analysis does not capture the true strength, deformability, and damping of a structure. R, C_d, and Ω_0 factors are used to mitigate the shortcomings of an elastic analysis, but these factors are arbitrary. Collapse-proof seismic design should be carried out by using true capacities and true demands without the use of R, C_d, and Ω_0 factors.

References

AISC (American Institute of Steel Construction). 2016. *Seismic provisions for structural steel buildings.* AISC 341-16. Chicago: AISC.

ASCE. 2022. *Minimum design loads for building and other structures.* ASCE 7-22, Reston, VA: ASCE.

CSI (Computers and Structures, Inc.). 2011. *SAP 2000.* Berkeley, CA: CSI.

Malhotra, P. K. 2005. "Return period of design ground motions." *Seismol. Res. Lett.* 76 (6): 693–699.

CHAPTER 9

Concluding Remarks

A ground motion is defined by its acceleration–deformation and cyclic-demand response spectra (ADRS and CDRS) for various values of damping. The size of an ADRS is captured by energy demand, *ED*; its shape is captured by energy centroid, T_{cg}, and normalized velocity, PGV_n. There is no single parameter that completely defines the strength of ground shaking, but *ED* appears to be most suited for engineered structures. Energy demand correlates well with the square of peak ground velocity *PGV*.

The importance of peak ground acceleration *PGA* should not be overemphasized. *PGA* is a measure of high frequencies that are easily repelled by ductile structures. Only stiff and brittle structures respond to *PGA*. Whether it is a recorded ground motion or a simulated ground motion (for shake-table testing), *ED* (or *PGV*) are more important than *PGA* in predicting the response of ductile structures.

From the perspective of collapse prevention, the important characteristics of a structure are its seismic toughness, damping at ultimate deflection, and cyclic capacity. Seismic toughness and damping can be rolled into the minimum energy demand needed to collapse the structure. Cyclic capacity is not sufficiently clarified in current seismic guidelines. In this book, conservative estimates of cyclic capacity have been made. There is a need to define plastic rotation/elongation limits for different number of cycles: one, two, three, four, and so on.

The strongest ground motions recorded so far have a negligible chance of exceedance compared to the chance of collapse of structures per current design standards.

It is practical to design structures made of ordinary materials and established technologies to avoid collapses during the strongest recorded ground motions. Nonstructural systems could be designed to accommodate drifts, and yielded structural elements could be replaced after a damaging earthquake. Such designs can be standardized for worldwide applications without the need for site-specific ground motion studies. In many parts of the world, local codes and site-specific studies are unreliable because of the lack of sufficient geosciences data.

Seismic design based on elastic period and 5% damping is arbitrary. The period and damping of a structure are not fixed – they change during the ground shaking. A nonlinear analysis is needed to calculate the response of structures during strongest ground motions. Nonlinear-static analysis described in this book is generally more conservative than a nonlinear-dynamic analysis. Nonlinear-dynamic analysis should always be preceded by a nonlinear-static analysis. The model for nonlinear-dynamic analysis should be simplified to minimize numerical errors.

Probabilistic seismic hazard analysis appears to overpredict the extreme ground motions. Overprediction of hazard is not helpful because it results in (1) increased cost of insurance, (2) cancellation of useful projects, (3) inefficient allocation of resources, and (4) a less-serious response to the predicted hazard.

It is difficult to decide acceptable risk for critical structures such as hospitals, emergency shelters, large dams, nuclear reactors, and nuclear waste storage facilities. It is proposed that critical structures should be designed to withstand the strongest recorded ground motions. Structures whose collapse can cause many deaths should also be designed to withstand the strongest ground motions. Collapse of numerous small structures can be as devastating as the collapse of one large structure. The aggregate effect of earthquakes should not be overlooked.

It is much more reassuring to say that "a structure has been designed to withstand the strongest ground motion," than to say that "a structure has been designed to withstand two-thirds of 2,475-year MRP ground motion reduced by risk factors, deterministic limit, and R factor."

Elastic analysis does not capture the true strength, deformability, and damping of a structure. R, C_d, and Ω_0 factors are used to mitigate the shortcomings of an elastic analysis, but these factors are arbitrary. They can vary from one building to another even when the buildings have the same lateral load resisting system. The R factor is also very sensitive to the frequency composition of the ground motion. Collapse-proof seismic design should be carried out by using true capacities and true demands without the use of R, C_d, and Ω_0 factors.

The physics of the Earth is too complex to be captured by any models soon. Therefore, predictions of future ground motions will remain highly subjective and uncertain for a long time. When uncertainty is not an option, structures should be designed to the strongest recorded ground motions.

Damping of Hysteretic System

The equation for equivalent-viscous damping of a hysteretic system is derived in this appendix.

For $D \leq D_e$, the system is elastic, and its damping is viscous in nature. The elastic damping as a fraction of critical damping is denoted as ζ_e. For $D > D_e$, the effect of viscous damping reduces, but the loss of viscous damping is made up by increase of hysteretic damping. Figure A-1 shows the cyclic force–deformation relationship of a hysteretic system undergoing peak deformation $D > D_e$. The energy dissipated during a cycle of peak deformation D is given by the following equation:

$$E_D = 2\pi^2 \frac{C_e}{T_{ef}} D_e^2 + 4F(D - D_e) \cdot \beta \tag{A-1}$$

In the previous equation, the first term on the right side is energy loss per cycle due to viscous damping [Equation (1.5) in Malhotra 2021] and the second term on the right side is energy loss per cycle due to hysteresis. Note that β is the ratio between the area of a realistic hysteresis loop and that of an

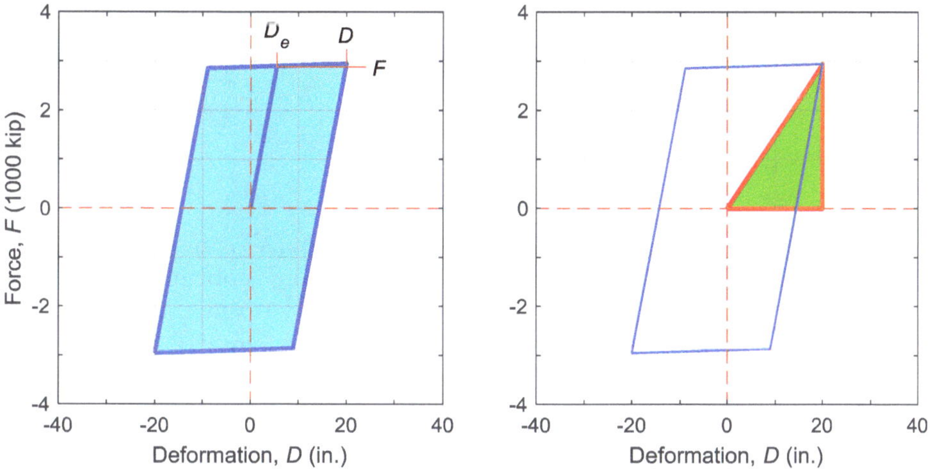

Figure A-1. Cyclic-force deformation relationship showing hysteretic energy (left) and strain energy (right).

ideal hysteresis loop; it is discussed in Section 3.5. C_e, the elastic damping constant, is given by the following expression:

$$C_e = 2\zeta_e \frac{2\pi}{T_e} m \tag{A-2}$$

where T_e is the elastic period and m is the mass of the system. In Equation (A-1), T_{ef} = the effective period of the system undergoing deformation D. For $D > D_e$, it is given by the following expression:

$$T_{ef} \approx T_e \sqrt{\frac{D}{D_e}} \tag{A-3}$$

The elastic period T_e is given by the expression

$$T_e = 2\pi \sqrt{\frac{m}{k_e}} \tag{A-4}$$

Substituting Equations (A-2), (A-3), and (A-4) into Equation (A-1) gives

$$E_D = 2\pi\zeta_e k_e D_e^2 \sqrt{\frac{D_e}{D}} + 4F(D - D_e) \cdot \beta \tag{A-5}$$

The "strain" energy is the area of the triangle in Figure A-1; it is given by the expression

$$E_s = \frac{1}{2} F \left(1 + \alpha \left(\frac{D}{D_e} - 1\right)\right) D \tag{A-6}$$

in which α is the post-/pre-yield stiffness of the system after accounting for strain hardening and P-Δ. It is discussed in Section 3.5. The equivalent viscous damping (Chopra 2017) is given by the expression

$$\zeta_{eq} = \frac{1}{4\pi} \frac{E_D}{E_s} \tag{A-7}$$

Substituting Equations (A-5) and (A-6) into Equation (A-7) and assuming that $F \approx k_e D_e$, the following expression for equivalent damping is obtained:

$$\zeta_{eq} = \zeta_e \qquad \text{for } D \leq D_e$$

$$\zeta_{eq} = \zeta_e \left(\frac{D_e}{D}\right)^{3/2} + \frac{2\beta}{\pi} \frac{\left(1 - \frac{D_e}{D}\right)}{\left(1 + \alpha\left(\frac{D}{D_e} - 1\right)\right)} \qquad \text{for } D > D_e \tag{A-8}$$

The previous equation gives the equivalent damping for a cycle of amplitude D. During the earthquake, the structure experiences cycles of various amplitudes between 0 and D. The average damping for cycles of amplitudes between 0 and D is given by the expression

$$\zeta = \frac{1}{D}\int \zeta_{eq} \cdot dD \tag{A-9}$$

After substituting ζ_{eq} from Equation (A-8) into Equation (A-9) and solving the integral, the following expression for damping is obtained:

$$\zeta = \zeta_e \qquad \text{for } D \leq D_e$$

$$\zeta = \frac{\zeta_e}{D/D_e}\left(3 - \frac{2}{\sqrt{D/D_e}}\right) + \frac{2\beta}{\pi\alpha(1-\alpha)(D/D_e)}\log_n\left(\frac{1-\alpha+\alpha D/D_e}{(D/D_e)^{\alpha}}\right) \qquad \text{for } D > D_e \tag{A-10}$$

For $\alpha = 0$, the previous equation reduces to

$$\zeta = \zeta_e \qquad \text{for } D \leq D_e$$

$$\zeta = \frac{\zeta_e}{D/D_e}\left(3 - \frac{2}{\sqrt{D/D_e}}\right) + \frac{2\beta}{\pi}\left(1 - \frac{1+\log_n(D/D_e)}{D/D_e}\right) \qquad \text{for } D > D_e \tag{A-11}$$

The first term on the right side is viscous damping and the second term is hysteretic damping. As D increases, the first term decreases but the second term increases.

References

Chopra, A. K. 2017. *Dynamics of structures.* 5th ed. London: Pearson Publication.

Malhotra, P. K. 2021. "Ground motions from past earthquakes." Chap. 1 in *Seismic analysis of structures and equipment.* Cham, Switzerland: Springer. https://doi.org/10.1007/978-3-030-57858-9_1.

Index

Figures are indicated by *f*; tables are indicated by *t*.

acceleration: histories, 1, 2*f*; Los Gatos Presentation Center, 87, 87*f*; maximum-horizontal, 1–2; NIG019 ground motion, 115, 115*f*; resultant horizontal, 2–3, 3*f*; sensors, 1; Sylmar ground motion, 94, 94*f*; Takatori ground motion, 101, 101*f*; Tarzana ground motion, 83, 84*f*; TCU052 ground motion, 111, 112*f*; TCU065 ground motion, 90, 91*f*; TCU068 ground motion, 118, 119*f*; TCU084 ground motion, 104, 105*f*, 122*f*; TK3123 ground motion, 108, 108*f*; "true," 10

acceleration–deformation response spectrum (ADRS), 12–15, 145; collapse-inducing ground motion, 72, 72*f*–73*f*, 76, 76*f*; for damping, 15, 16*f*; envelope ground motion, 138, 139*f*, 142, 143*f*; 5% damping, 76, 76*f*; logarithmic, 12, 125, 127*f*; mean return period (MRP), 24–25, 24*f*–25*f*; TAPS ground motion, 13*f*–14*f*, 55, 56*f*, 137, 137*f*, 140, 141*f*; 3-story steel SMF building, weakest ground motion, 71–72, 72*f*, 74, 74*t*, 79*f*; 20-story steel EBF building, weakest ground motion, 76, 76*f*–77*f*

accelerometers, 1

ADRS. *See* acceleration–deformation response spectrum (ADRS)

aftershocks, effect of, 121–123

AISC 341-16 standard, 41, 45, 50; cyclic test protocol, 51*f*–52*f*

ASCE 7-22, 29, 135–136

ASTM A992 steel, 29, 42, 136, 140

black curve: CHY080 ground motion, 98–99; Los Gatos Presentation Center, 88–89; NIG019 ground motion, 115, 117; Sylmar ground motion, 95, 97; Takatori ground motion, 102–103; Tarzana ground motion, 84, 86; TCU052 ground motion, 112–113; TCU065 ground motion, 91, 93; TCU068 ground motion, 118, 120; TCU084 ground motion, 105–106; TK3123 ground motion, 109–110

buckling-restrained brace (BRB), 128–129

buckling-restrained braced frame (BRBF) building, 128

California Department of Water and Power (DWP), 94

California Geological Survey (CGS), 83

capacity curve, 129–131, 130*f*; SSI, 131, 131*f*; for 3-story building, 35–37, 35*f*, 50*f*; for 20-story building, 46–48, 48*f*, 50*f*

CDRS. *See* cyclic-demand response spectra (CDRS)

center of gravity (CG), 29–30; building deflection, 58; 20-story steel EBF building, 65

Chi-Chi, Taiwan Earthquake: CHY080 station, 97–100; TCU052 station, 111–114; TCU065 station, 90–94; TCU068 station, 118–121; TCU084 station, 104–107, 122*t*

collapse-inducing ground motion: ADRS, 72, 72*f*–73*f*, 76, 76*f*; deformability effect, 78, 78*f*, 78*t*; hysteresis effect, 79–80, 80*f*, 80*t*; weakest, 71–78

collapse-inducing response spectrum, 75*f*

collapse prevention, 80

collapse-proof seismic design, 125; capacity and damping curves, 129–131; envelope response spectra, 125, 126*f*–127*f*; preliminary design, 128–129; *R*, C_d, and $\grave{U}_0$ factors, 135–144; SSI, 131, 131*f*

collapse-proof structures, 125

cyclic capacity, 50–52

cyclic-demand response spectra (CDRS), 19–20, 20*f*, 145; envelope for damping, 125, 127*f*; TAPS ground motion, 55, 56*f*

cyclic force–deformation relationship, 147, 147*f*

damping, 145. *See also* 5% damping; hysteretic system, 147–149, 147*f*; ratio, 6

damping-adjusted demand curve, 72, 73*f*, 76, 77*f*

damping curve: ideal *versus* realistic hysteresis loop, 38, 38*f*; for 3-story building, 37–38, 39*f*, 51*f*, 57*f*; for 20-story building, 48–49, 49*f*, 51*f*, 63*f*

damping curves, 129–131, 130*f*; SSI, 131, 132*f*

deflection amplification factor (C_d), 135–136, 140

deformability, 128–129; effect, 78, 79*t*; 3-story steel SMF building, 139, 139*t*; 20-story steel EBF building, 142, 143*t*

deformation response spectrum, 7–9, 8*f*–9*f*

demand-capacity equilibrium, 58, 59*f*, 63, 64*f*, 72, 74*f*, 76, 78*f*; CHY080 ground motion, 98–99, 99*f*–100*f*; Los Gatos Presentation Center, 88, 88*f*, 90*f*; NIG019 ground motion, 115, 116*f*–117*f*, 117; Sylmar ground motion, 95–96, 95*f*–96*f*; Takatori ground motion, 102–103, 102*f*–103*f*; Tarzana ground motion, 84–85, 85*f*–86*f*; TCU052 ground motion, 112–113, 112*f*, 114*f*; TCU065 ground motion, 93, 93*f*; TCU068 ground motion, 118, 119*f*, 120, 121*f*; TCU084 ground motion, 105–106, 106*f*–107*f*; TK3123 ground motion, 109–110, 109*f*–110*f*

demand curve: CHY080 ground motion, 98–99; damping-adjusted, 72, 73*f*; Los Gatos Presentation Center, 88–89; NIG019 ground motion, 115, 117; Sylmar ground motion, 95–97; Takatori ground motion, 102; TAPS ground motion, 55, 57, 58*f*; Tarzana ground motion, 84–86; TCU052 ground motion, 112–113; TCU065 ground motion, 91, 93; TCU068 ground motion, 118, 120; TCU084 ground motion, 105–106; for 3-story steel SMF building, 55–57; TK3123 ground motion, 109–110; for 20-story EBF building, 62–63

Disaster and Emergency Management Presidency of Türkiye (AFAD), 108

displacement: histories, 3–4, 4*f*; Los Gatos Presentation Center, 87*f*; NIG019 ground motion, 115, 115*f*; resultant horizontal, 5, 5*f*; Sylmar ground motion, 94*f*; Takatori ground motion, 101*f*; Tarzana ground motion, 84*f*; TCU052 ground motion, 112*f*; TCU065 ground motion, 91*f*; TCU068 ground motion, 118, 119*f*; TCU084 ground motion, 105*f*, 122*f*; TK3123 ground motion, 108*f*

drift ratio: histories of, 60, 60*f*; inter-story, 60; in N72E direction, 61, 61*f*; plastic, 39, 41; resultant, 61, 61*f*; ultimate, 32, 44

ductility ratio, 32, 44, 49–50, 125

eccentrically braced frame (EBF) system, 29, 39. *See also* 20-story steel EBF building; in flexure and shear, 40*f*; geometry, 41; plastic shears in, 43*f*; 20-story building, 39–42, 41*f*; two-story-X, 41–42

ED. *See* energy demand (ED)

elastic analysis, 146
elastic stiffness, 31, 44
elastic strength, 142, 143*t*
energy centroid, 16, 76; TAPS ground motion, 15*f*
energy demand (ED), 14–16, 71, 145; CHY080 ground motion, 97–99; LGPC ground motion, 88–89; NIG019 ground motion, 115, 117; recorded ground motions, 17*f*; reduction in, 57; Sylmar ground motion, 95–96; Takatori ground motion, 101, 103; TAPS ground motion, 15*f*; Tarzana ground motion, 84–85; TCU052 ground motion, 111, 113; TCU065 ground motion, 91–92; TCU068 ground motion, 118, 120; TCU084 ground motion, 104–106; TK3123 ground motion, 108, 110
equation of motion (EOM), 7, 55

5% damping, 15; ADRS, 76, 76*f*; linear plot of 5%, 14, 14*f*; logarithmic plot of 5%, 13*f*; response spectrum, ADRS, 72, 72*f*; tripartite plot of 5%, 11, 13*f*
force–deformation relationship, 37, 37*f*, 129

ground motions, ix–x. *See also* strongest ground motions; central period, 5; collapse-inducing, 71–81; cyclic demand, 18–20; for engineered structures, 17–18, 17*t*; Hickman, Arkansas, 25, 25*f*; histories, 1–6; indirect parameters, 20–21; narrow-banded, 71, 76; predicted *versus* strongest recorded, 23–26; pulse-like, 17; response spectrum of, 75*f*; in San Francisco, California, 23–24, 24*f*; strongest recorded, 16–18, 18*t*; TAPS. *See* TAPS ground motion; weakest collapse-inducing, 71–78

Hickman, Arkansas, ground motion, 25, 25*f*
hysteresis, 37; on collapse-inducing ground motion, 79–80, 80*f*; ideal *versus* realistic loop, 38, 38*f*, 45; on weakest ground motion, 80, 80*t*

ideal hysteresis loop, 38, 38*f*, 45

Joyner–Boore distance, 18

Kobe, Takatori Earthquake, 101–104

linear-elastic SDOF system, 6
link-beams, plastic-shear capacities, 42, 42*t*
logarithmic ADRS, 12, 125, 127*f*
Loma Prieta, Los Gatos Presentation Center, 87–90

magenta curve: CHY080 ground motion, 98–99; Los Gatos Presentation Center, 88–89; NIG019 ground motion, 115, 117; Sylmar ground motion, 95–97; Takatori ground motion, 102–103; Tarzana ground motion, 84–86; TCU052 ground motion, 112–113; TCU065 ground motion, 91, 93; TCU068 ground motion, 118, 120; TCU084 ground motion, 105–106; TK3123 ground motion, 109–110
mean return period (MRP), 24–25, 24*f*–25*f*, 135

narrow-banded ground motions, 71, 76
natural period, 8, 35
Niigata, Japan, NIG019 Earthquake, 114–118
nonlinear analysis, 145
nonlinear-dynamic analysis, 55, 145–146; 3-story steel SMF building, 59–61; 20-story EBF building, 66–67

nonlinear-static analysis, 55; 3-story steel SMF building, 58–59; 20-story EBF building, 63–66
nonlinear system, 14
nonstructural systems, 145
Northridge, California Earthquake: Sylmar station, 94–97; Tarzana station, 83–87
Nurdagi, Türkiye, TK3123 Earthquake, 108–111

overstrength factor (Ω_0), 135–136, 140

Pacific Earthquake Engineering Research Center (PEER), 87, 90, 97, 101, 104, 111, 118
P-Δ effect, 33, 34*f*, 35–36, 45–46, 47*f*, 136
peak deformation (PD), 7–9, 57, 137, 141
peak force (PF), 9–10
peak ground acceleration (PGA), 145
peak pseudo-acceleration (PPA), 9–10, 57
peak pseudo-velocity (PPV), 11
peak strain energy (PSE), 11
PEER. *See* Pacific Earthquake Engineering Research Center (PEER)
plastic drift ratio, 39, 41
plastic-shear capacity, 42, 42*t*
probabilistic seismic hazard analysis (PSHA), 23, 26
pseudo-acceleration response spectrum, 9–10, 10*f*
pseudo-velocity response spectrum, 11, 12*f*
pulse-like ground motion, 17
pushover curve: ideal *versus* realistic, 33*f*, 45*f*; P-Δ effect, 33, 34*f*, 35–36, 45–46, 47*f*; strain hardening, 33, 34*f*, 45, 46*f*–47*f*; for 3-story building, 30–35, 32*f*; for 20-story building, 42–46, 44*f*

realistic hysteresis loop, 38, 38*f*, 45
response modification factor *(R)*, 135, 137–142, 138*t*, 146
response spectrum:
acceleration–deformation, 12–15; collapse-inducing, 75*f*; deformation, 7–9, 8*f*–9*f*; envelope, 125, 126*f*–127*f*; of ground motion, 75*f*; pseudo-acceleration, 9–10, 10*f*; pseudo-velocity, 11, 12*f*; tripartite plot of, 11–12, 13*f*
resultant drift ratio, 61, 61*f*
rigid offsets, 30

safety factor (SF), 138–139, 139*t*, 141–142, 143*t*
San Francisco, California ground motion, 23–24, 24*f*
SDOF. *See* single-degree-of-freedom (SDOF) systems
seismic design, 145
seismic toughness (ST), 36, 125, 125*t*, 129, 131, 145; 3-story steel SMF building, 36, 36*f*; 20-story steel EBF building, 48, 48*f*
SF. *See* safety factor (SF)
single-degree-of-freedom (SDOF) systems, 6; deformations of, 7, 8*f*, 19*f*; free-vibration responses, 7*f*; structure of, 6*f*
SMF. *See* special moment frame (SMF)
soil–structure interaction (SSI), 131, 131*f*–132*f*
special moment frame (SMF), 29, 39. *See also* 3-story steel SMF building; plastic moments in, 30–31, 31*f*
spring force, 9
SSI. *See* soil–structure interaction (SSI)
ST. *See* seismic toughness (ST)
strain hardening, 33, 34*f*, 45, 46*f*–47*f*
strongest ground motions, 145; aftershocks, damage from, 121–123; Chi-Chi, Taiwan, 90–94, 97–100, 104–107, 111–114, 118–121; Kobe,

Takatori, 101–104; Loma Prieta, Los Gatos Presentation Center, 87–90; Niigata, Japan, NIG019, 114–118; Northridge, Sylmar, 94–97; Northridge, Tarzana, 83–87; Nurdagi, Türkiye, TK3123, 108–111; recorded, 16–18, 18*t*

TAPS ground motion: acceleration at, 1–2, 2*f*; acceleration–deformation response spectrum (ADRS), 13*f*–14*f*, 55, 56*f*, 137, 137*f*, 140, 141*f*; cyclic-demand response spectra (CDRS), 19, 20*f*, 55, 56*f*; deformation response spectrum, 8, 9*f*; demand curve, 55, 57, 58*f*, 64*f*; displacements, 4, 4*f*; energy demand (ED), 14–15, 15*f*; horizontal accelerations, 3*f*; horizontal displacements, 5, 5*f*; horizontal velocities, 3–4, 4*f*; N72E direction on, 61, 61*f*; pseudo-acceleration response spectrum, 10, 10*f*; pseudo-velocity response spectrum, 11, 12*f*; roof deflection, 3-story building, 61, 62*f*; 3-story steel SMF building, 59*t*, 60*f*–61*f*; 20-story steel EBF building, 65*t*; velocity in, 3, 3*f*

3-story buildings. *See also* 3-story steel SMF building: *versus* 20-story buildings, 49–50

3-story steel BRBF building, 128, 128*f*

3-story steel SMF building, 29–30, 30*f*; ADRS weakest ground motion, 71–72, 72*f*, 74, 74*t*, 79*f*; capacity curve for, 35–37, 35*f*, 50*f*; to CHY080 ground motion, 97–98, 99*f*, 99*t*; cyclic force–deformation relationship, 37, 37*f*; damage suffered by, 62; damping, 139, 139*t*; damping-adjusted demand curve, 72, 73*f*; damping curve for, 37–38, 39*f*, 51*f*; deflection amplification factor (C_d), 136; deformability, 78, 79*t*, 139, 139*t*; demand-capacity equilibrium, 58, 59*f*, 72, 74*f*; demand curve for, 55–57; energy diagram, 72, 74*f*; energy repelled by, 59, 60*f*; hysteresis on, 80, 80*t*; to Los Gatos Presentation Center, 88–89, 88*f*, 89*t*; to NIG019 ground motion, 115–116, 116*f*, 116*t*; nonlinear-dynamic analysis, 59–61; nonlinear-static analysis, 58–59; overstrength factor (Ω_0), 136; pushover curve for, 30–35, 32*f*; response modification factor *(R)*, 137–139, 138*t*; roof deflection on TAPS, 61, 62*f*; safety factor, 139, 139*t*; seismic toughness (ST), 36, 36*f*; to Sylmar ground motion, 95–96, 95*f*, 96*t*; to Takatori ground motion, 101–103, 102*f*, 102*t*; on TAPS ground motion, 59*t*, 60*f*–61*f*; to Tarzana ground motion, 84, 85*f*, 85*t*; to TCU052 ground motion, 111–113, 112*f*, 113*t*; to TCU065 ground motion, 91–92, 92*f*, 92*t*; to TCU068 ground motion, 118–120, 119*f*, 119*t*; to TCU084 ground motion, 104–105, 106*f*, 106*t*; to TK3123 ground motion, 108–109, 109*f*, 110*t*; true (inelastic) strength, 138–139, 139*t*

Trans Alaskan Pump Station 10 (TAPS 10), 1

tripartite response spectrum, 11–12, 13*f*

true (inelastic) strength, 138–139, 139*t*

20-story buildings. *See also* 20-story steel EBF building: *versus* 3-story buildings, 49–50

20-story steel EBF building, 29, 39–42; ADRS weakest ground motion, 76, 76*f*–77*f*; capacity curve for, 46–48, 47*f*, 50*f*; to CHY080 ground motion, 98–100, 100*f*, 100*t*; damage suffered by, 67–68; damping, 142, 143*t*; damping-adjusted demand curve, 76, 77*f*; damping curve for, 48–49, 49*f*, 51*f*, 63*f*; deflection amplification factor (C_d), 140; deformability, 142, 143*t*; demand-capacity equilibrium, 76, 78*f*; demand curve for, 62–63; elastic strength, 142, 143*t*; energy

diagram for, 76, 78*f*; to Los Gatos Presentation Center, 89, 90*f*, 90*t*; to NIG019 ground motion, 117–118, 117*f*, 118*t*; nonlinear-dynamic analysis, 66–67; nonlinear-static analysis, 63–66; overstrength factor (Ω_0), 140; pushover curve for, 42–46, 44*f*; response modification factor *(R)*, 140–142, 142*t*; safety factor, 142, 143*t*; seismic toughness (ST), 48, 48*f*; to Sylmar ground motion, 96–97, 96*f*, 97*t*; to Takatori ground motion, 103–104, 103*f*, 104*t*; on TAPS ground motion, 65*t*; to Tarzana ground motion, 85–87, 86*f*, 86*t*; to TCU052 ground motion, 113–114, 114*f*, 114*t*; to TCU065 ground motion, 92–93, 93*f*, 94*t*; to TCU068 ground motion, 120–121, 121*f*, 121*t*; to TCU084 ground motion, 105–107, 107*f*, 107*t*; to TK3123 ground motion, 110–111, 110*f*, 111*t*; weakest ground motion, 74, 74*t*
two-story-X EBF, 41–42

ultimate drift ratio, 32, 44
US Geological Survey (USGS), 23–25

velocity: ground, 3; histories, 2–3, 3*f*; horizontal, 3–4, 4*f*; Los Gatos Presentation Center, 87*f*; NIG019 ground motion, 115, 115*f*; Sylmar ground motion, 94*f*; Takatori ground motion, 101*f*; Tarzana ground motion, 84*f*; TCU052 ground motion, 112*f*; TCU065 ground motion, 91*f*; TCU068 ground motion, 118, 119*f*; TCU084 ground motion, 105*f*, 122*f*; TK3123 ground motion, 108*f*
viscous damping, 55

weakest collapse-inducing ground motion, 71–72; deformability on 3-story building, 78, 79*t*; hysteresis on 3-story building, 80, 80*t*; 3-story steel SMF building, 74, 74*t*; 20-story steel EBF building, 74, 74*t*
Wen-Bouc model, 59

About the Author

Dr. Praveen K. Malhotra, P.E., is an expert on seismic ground motions and their effect on structures. After receiving his education from IIT Delhi, AIT Bangkok, and Rice University, he worked alongside geologists, seismologists, engineers, and insurers for many years and founded his own consulting company, StrongMotions Inc. He is the author of two previous books and numerous peer-reviewed research papers. His first book, *Seismic Analysis of Structures and Equipment*, presents an innovative approach to seismic design. His second book, *Site-Specific Ground Motions for Seismic Design of Buildings and Other Structures*, was published by ASCE Press in 2022. Both books describe complex ideas in an intuitive manner without compromising depth. He is an inventor, innovator, and trainer for seismic topics for ASCE in the United States and abroad.

FSC
www.fsc.org
MIX
Papier | Fördert gute Waldnutzung
FSC® C083411

Zeitfracht Medien GmbH
Ferdinand-Jühlke-Straße 7
99095 Erfurt, Deutschland
produktsicherheit@kolibri360.de